Lecture Notes in Geosystems Mathematics and Computing

Series Editors
Willi Freeden, Technische Universität Kaiserslautern, Kaiserslautern, Germany
M. Zuhair Nashed, University of Central Florida, Orlando, FL, USA
Otmar Scherzer, University of Vienna, Vienna, Austria

Birkhäuser

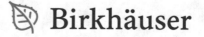

More information about this series at http://www.springer.com/series/15481

Andrea Aspri

An Elastic Model for Volcanology

Birkhäuser

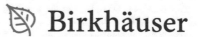

Andrea Aspri
Austrian Academy of Sciences
Johann Radon Institute for Computational
Linz, Austria

Lecture Notes in Geosystems Mathematics and Computing
ISBN 978-3-030-31474-3 ISBN 978-3-030-31475-0 (eBook)
https://doi.org/10.1007/978-3-030-31475-0

Mathematics Subject Classification: 35C20, 35R30, 35J57, 74B05, 86A60, 31B10, 35C15, 86A22

This book is published under the imprint Birkhäuser, www.birkhauser-science.com by the registered company Springer Nature Switzerland AG.
The registered company address is: Gewerbestrasse 11, 6330 Cham, Switzerland

Preface

This monograph focuses on a rigorous mathematical study of a linear elastic model applied in geophysics to explain deformation effects generated by inflating or deflating magma chambers. The modeling assumptions translate, from a mathematical view point, into a Neumann boundary value problem for the Lamé system in a half-space (representing the earth's crust) and an embedded pressurized cavity (depicting the magma chamber), where the pressure is supposed to be constant. This boundary value problem is well known in the geophysics literature. For instance, one can consult [111] and [125] where the case of a small spherical magma chamber is analysed. In particular, McTigue in [111] provides an asymptotic expansion at second order of the solution of the problem when the radius of the sphere goes to zero. Here, we present the results which are obtained generalizing the spherical geometry of the cavity to the case of a generic bounded Lipschitz domain, contained in a half-space. In a rigorous mathematical framework, we study the well-posedness of the elastic model (the direct problem) utilizing two different analytical techniques: One is based on a variational formulation which uses weighted Sobolev spaces. The other one uses single and double layer potentials which corresponds to rewrite the differential problem in terms of integral equations. Employing two different paths to prove the well-posedness of the elastic model is due to the tasks which we have in mind: On one hand, assuming the cavity to be small, we prove an asymptotic expansion at first order, of the analytical solution, as the diameter of the cavity goes to zero. To deal with this problem, we use the approach of Ammari and Kang, see for example [21, 27], based on layer potential techniques. On the other hand, we analyse the following inverse problem: Are we able to detect uniquely and in a stable way the finite pressurized cavity from partial measurements of the displacement field on the boundary of the half space when the pressure on the boundary of the cavity is known? For this second task the variational approach based on weighted Sobolev spaces is more powerful than the integral one. For the uniqueness issue and to determine stability estimates for the inverse problem, we follow the approach of [115, 116].

We mention that the analysis of the elastic boundary value problem is preceded by that of a simplified scalar problem, based on the Laplace equation. As far as we know, this scalar version doesn't have a real physical application but it serves for the non-expert reader to clarify the phylosophy behind the use of the some mathematical tools, before coming across the technicalities of the elastic theory.

Essentially this monograph originates from my Ph.D. thesis which I wrote during my Ph.D. program in Mathematics, from 2013 to 2017, at "Sapienza", University of Rome. This book would have not been possible without the support, suggestions, and the hard work of my two advisors Corrado Mascia and Elena Beretta and without the outstanding collaboration with Edi Rosset. I feel indebted to all of them for their help and encouragement. I would also like to thank Maurizio Battaglia and Corrado Mascia again for bringing my attention to this geophysical problem. Moreover I want to thank Cherif Amrouche and Habib Ammari for providing me with some useful papers and helpful advices. Finally, a special thanks goes to Otmar Scherzer which read my Ph.D. thesis and encouraged me to write this book.

Linz, Austria Andrea Aspri

Contents

Table of Notations

$\boldsymbol{x}, \boldsymbol{y}, \boldsymbol{z}$	Points in $\mathbb{R}^d$, with $d \geq 3$
$\boldsymbol{x}'$	First $d-1$ components of the point $\boldsymbol{x}$ that is $\boldsymbol{x}' = (x_1, \cdots, x_{d-1})$
$\widetilde{\boldsymbol{x}}$	Given a point $x \in \mathbb{R}^d_-$, $\widetilde{x}$ represents the reflected point $(\boldsymbol{x}', -x_d)$
$\mathbb{R}^d_-$	Denotes the half-space $\{\boldsymbol{x} = (x_1, \cdots, x_d) \in \mathbb{R}^d : x_d < 0\}$
$\mathbb{R}^{d-1}$	Boundary of the half-space $\mathbb{R}^d_-$
$B_r(\boldsymbol{x})$	Denotes the d-dimensional ball with centre $\boldsymbol{x}$ and radius $r > 0$
$B_r^-(\boldsymbol{x})$	It is the half ball $B_r^-(\boldsymbol{x}) = \{\boldsymbol{y} : \|\boldsymbol{y} - \boldsymbol{x}\| < r, \ \ y_d - x_d < 0\}$
Ω	Bounded Lipschitz domain in $\mathbb{R}^d$
ω_d	Area of the $(d-1)$-dimensional unit sphere
$\boldsymbol{u}, \boldsymbol{v}, \boldsymbol{w}, \ldots$	Vectors in $\mathbb{R}^d$
$\boldsymbol{n}$	Unit outer normal vector to a surface
$\boldsymbol{u} \cdot \boldsymbol{v}$	Inner product between vectors $\boldsymbol{u}$ and $\boldsymbol{v}$
$\boldsymbol{u} \times \boldsymbol{v}$	Cross vector between $\boldsymbol{u}$ and $\boldsymbol{v}$
$\boldsymbol{u} \otimes \boldsymbol{v}$	Tensor product between vectors $\boldsymbol{u}$ and $\boldsymbol{v}$
$\mathbf{A}, \mathbf{B}, \ldots$	Matrices and second-order tensors
$\mathbf{I}$	Identity matrix
$\mathbf{A}^T$	Transpose of the matrix $\mathbf{A}$
$\widehat{\mathbf{A}}$	Symmetric part of the matrix $\mathbf{A}$, that is $\widehat{\mathbf{A}} = \frac{1}{2}\left(\mathbf{A} + \mathbf{A}^T\right)$
$\mathbf{A} : \mathbf{B}$	Inner product between the two matrices $\mathbf{A}$ and $\mathbf{B}$ that is $\mathbf{A} : \mathbf{B} = \sum_{i,j} a_{ij} b_{ij}$
$\|\mathbf{A}\|$	Norm induced by the matrix inner product, that is $\|\mathbf{A}\| = \sqrt{\mathbf{A} : \mathbf{A}}$
I	In Chap. 2 it represents the identity map
$\Gamma(\boldsymbol{x})$	Fundamental solution of the Laplace operator
$S_\Omega\varphi(\boldsymbol{x})$	Single layer potential for the Laplace operator with moment φ
$D_\Omega\varphi(\boldsymbol{x})$	Double layer potential for the Laplace operator with moment φ
$N(\boldsymbol{x}, \boldsymbol{y})$	Neumann function of the half-space for the Laplace operator
κ_d	Constant in the definition of Γ function, $\kappa_d := 1/\omega_d(2-d)$
$\mathbb{A}, \mathbb{B}, \ldots$	Fourth-order tensors
$\mathbb{C}$	Fourth-order elasticity tensor
$\mathbb{I}$	Fourth-order identity tensor such that $\mathbb{I}\mathbf{A} = \widehat{\mathbf{A}}$

μ, λ	Lamé parameters of the linear elasticy theory
ν	Poisson ratio. The identity $\nu = \lambda/2(\lambda + \mu)$ holds
$\mathcal{L}$	Elastostatic Lamé operator, that is $\mathcal{L}\boldsymbol{u} := \mu\Delta\boldsymbol{u} + (\lambda + \mu)\nabla\text{div}\,\boldsymbol{u}$
$\dfrac{\partial \boldsymbol{u}}{\partial \boldsymbol{\nu}}$	Conormal derivative, that is $\frac{\partial \boldsymbol{u}}{\partial \boldsymbol{\nu}} := (\mathbb{C}\widehat{\nabla}\boldsymbol{u})\boldsymbol{n} = \lambda(\text{div}\,\boldsymbol{u})\boldsymbol{n} + 2\mu(\widehat{\nabla}\boldsymbol{u})\boldsymbol{n}$
$\boldsymbol{\Gamma}(\boldsymbol{x})$	Fundamental solution of the Lamé operator (Kelvin-Somigliana matrix)
$\mathbf{N}(\boldsymbol{x}, \boldsymbol{y})$	Neumann function of the half-space related to the Lamé operator. $\mathbf{N}(\boldsymbol{x}, \boldsymbol{y}) = \boldsymbol{\Gamma}(\boldsymbol{x}, \boldsymbol{y}) + \mathbf{R}(\boldsymbol{x}, \boldsymbol{y})$, with $\mathbf{R}$ regular part
$\boldsymbol{N}^{(k)}(\boldsymbol{x}, \boldsymbol{y})$	k-Th column vector of the Neumann function $\mathbf{N}$
$\mathbf{S}^{\Gamma}\boldsymbol{\varphi}(\boldsymbol{x})$	Single layer potential related to the Lamé operator with kernel $\boldsymbol{\Gamma}$
$\mathbf{D}^{\Gamma}\boldsymbol{\varphi}(\boldsymbol{x})$	Double layer potential related to the Lamé operator with kernel $\boldsymbol{\Gamma}$
$\mathbf{S}^{R}\boldsymbol{\varphi}(\boldsymbol{x})$	Single layer potential with kernel $\mathbf{R}$
$\mathbf{D}^{R}\boldsymbol{\varphi}(\boldsymbol{x})$	Double layer potential with kernel $\mathbf{R}$
c_ν	Constant $c_\nu := 4(1 - \nu)(1 - 2\nu)$
c'_ν	Constant $c'_\nu := (1 - 2\nu)/(8\pi(1 - \nu))$
$C_{\mu,\nu}$	Constant $C_{\mu,\nu} := 1/(16\pi\mu(1 - \nu))$

Chapter 1
From the Physical to the Mathematical Model

Nowadays, collection of precise measurements of ground deformations is required to characterize the type of physical processes taking place inside the earth's crust and constrain the size and location of the source of unrest. This is, in fact, based on the assumption that tectonic and volcanic processes at depth are transmitted to the surface through the mechanical properties of the crust [43, 72, 125]. For this purpose, several geodetic techniques are employed such as leveling, tilt, GPS, and InSAR, see [38, 40–42, 125, 126]. All these measurements need to be interpreted by simplified mathematical models which play a key role in the explanation of how the crust responds to different physical phenomena and in distinguishing among possible sources of deformation [42, 43]. For instance, in monitoring active volcanoes, most of these models share the assumption that the crust can be thought, at least in a first approximation, as a homogeneous, isotropic elastic (flat) half-space, which contains the sources of unrest described by geometries of simple shape: a point dilatation [34, 114], a pressurized spherical source [111], a prolate spheroid [137], a horizontal penny-shaped source [82], or a tensile dislocation [121]. The use of these geometries represents certainly a great simplification but in the geological literature it is assumed that these models may reproduce the strain field created by storage areas and transport pathways, see [71]. Due to the complexity of Earth's crust, it is difficult to constrain the source of unrest by only considering measurements of ground deformations. In fact, monitoring of volcanoes activity is usually performed by combining different types of geophysical measurements: ground deformations, seismic swarms, and gravity changes. For more details see, for example, [41, 43, 44, 73, 125, 128] and the references therein. Modeling of the pattern and rate of displacement before and during eruptions can reveal much about the physics of active volcanoes [73]. This is especially true in studying stratovolcanoes or basaltic shield volcanoes, since their fast, short-term deformation is well associated with magma accumulation and eruptions, see [40, 41] and the references therein. Specifically, the monitoring of ground deformation has showed a cyclical volcanic activity of *inflation* and *deflation* period [125, 126]. When

A. Aspri, *An Elastic Model for Volcanology*, Lecture Notes in Geosystems Mathematics and Computing, https://doi.org/10.1007/978-3-030-31475-0_1

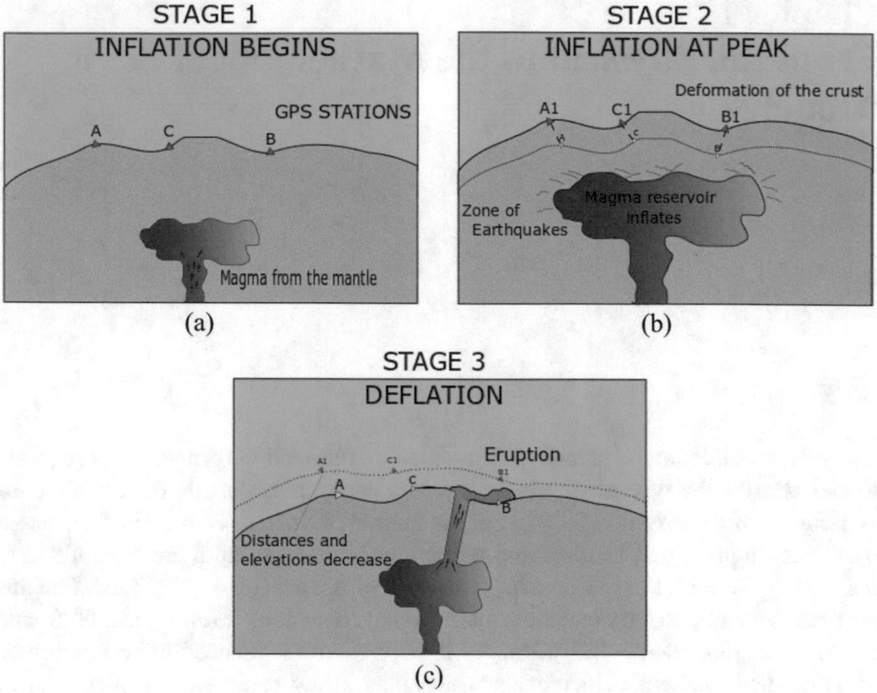

Fig. 1.1 Inflation-deflation cycle. (**a**) Magma comes from the mantle into the magma reservoir. (**b**) The inflation produces deformations. (**c**) Deflation period after an eruption

magma accumulates in crustal reservoirs, volcanoes inflate (see Fig. 1.1a, b). The observations indicate relatively long period of volcanic uplift. After that, rapid periods of subsidence follow. These deflation episodes are accompanied either by eruptions or by dike intrusion into the flanks of the volcano (see Fig. 1.1c).

Without being exhaustive, we can briefly explain and simplify the physical phenomenon in this way: as magma migrates toward the earth's surface, it forces aside and exerts stresses on the surrounding crust causing ground deformations and in some cases, since the crust is brittle, earthquakes. Consequently, the redistribution of the mass at depth generates changes in the material density producing as an effect of small anomalies in the gravity field. All these signals can be measured. However, since the subsurface structures beneath active volcanoes are extremely complex, the identification of the source of unrest is not straightforward. In fact, caldera unrest may be also caused by aqueous fluid intrusions, or interaction between the hydrothermal system and magma intrusions [52, 74, 134]. We highlight that the deformation measurements are sensitive only to changes in volume and pressure of the source; hence, they cannot provide information on the material density. Gravity measurements, however, can constrain the mass of the intrusion. Given the significant density difference between silicate melts (2500 kg/m^3) and hydrothermal fluids (1000 kg/m^3), it is reasonable to use density estimates from gravity measurements to distinguish between these two possible sources of caldera unrest [41].

In light of this, the main challenge is to interpret geodesy and gravity measurements jointly (see [41, 50, 122, 128]) in order to reach the following goals:

1. determine the geometry of subsurface magma bodies, i.e., whether the source of deformation is a dike, a roughly equidimensional chamber, or a hybrid source (mixture of different mantle sources);
2. to quantify parameters of the source, for example, its depth, dimensions, volume, density, and internal magma pressure [73, 125, 128].

To achieve these objectives a simplified/conceptual model has been conceived with a central magma chamber that is supplied with melt from the mantle. The pressure increases; hence, the ground is deformed producing gravity anomalies and deformation of the surface of the crust. After some time, the increasing pressure causes the fracture of the walls and a dike propagates carrying magma either to the surface or into the volcano flanks [125].

This is an extremely difficult inverse problem because modeling of surface deformations and gravities anomalies does not provide a unique description of the source of unrest [107] even in the ideal case of data without noise. Based on the elastic behavior of the Earth's crust, the ground deformations are interpreted in the framework of linear elasticity theory, see [38, 73, 126]. The gravity anomalies are interpreted by using potential theory, see [41] and reference therein.

In this book we will focus the attention on the mathematical analysis of the best-known and widely applied model in geology, i.e., the so-called Mogi model [114]. We first mention that the mathematical formulation of this model and its solution come back to Anderson [34] and later to Yamakawa [136]. Mogi was the first one to apply it to the study of ground deformation in a volcanic region due to inflating or deflating magma chambers. See, for example, description in [69, 107, 125]. Mogi model is based on many simplifications related to the geometry of the Earth's crust, of the magma chamber and its depth. Specifically, the magma chamber is described by a spherical cavity, with radius r_0 and depth d_0 such that $r_0 \ll d_0$, and it is assumed to be filled with an ideal incompressible fluid at equilibrium, so that the pressure p exerted on its boundary to the external medium is uniform. In addition, the surrounding crust is represented by an infinite, homogeneous, and isotropic elastic half-space, characterized by the Lamé parameters μ, λ. The crust boundary, which we can assume located on the plane $x_3 = 0$, is considered traction-free.

Under these assumptions, explicit expressions for the displacement vector fields on the boundary of the half-space can be provided. Explicitly, assuming that the center of the sphere is located at $\boldsymbol{z} = (z_1, z_2, z_3)$ with $z_3 < 0$, the solution of the Mogi model, i.e., the displacement $\boldsymbol{u} = (u_1, u_2, u_3)$ at a surface point $\boldsymbol{y} = (y_1, y_2, 0)$, is given by

$$u^\alpha(\boldsymbol{y}) = \frac{(1-\nu)}{\mu} \frac{\varepsilon^3 p(z_\alpha - y_\alpha)}{|\boldsymbol{z} - \boldsymbol{y}|^3}, \qquad u^3(\boldsymbol{y}) = -\frac{(1-\nu)}{\mu} \frac{\varepsilon^3 p\, d_0}{|\boldsymbol{z} - \boldsymbol{y}|^3} \tag{1.1}$$

as $\varepsilon := r_0/d_0 \to 0$, for $\alpha = 1, 2$, where $d_0 = -z_3$. We observe that the displacement field is rescaled by the product $\varepsilon^3 p$. This means that we can only determine this product and not either the volume or the pressure. In [111] McTigue proposed a formal second-order approximation of the solution in order to cover the case of a spherical cavity, with finite and small positive radius, and to potentially separate the contribution of the pressure and volume. He found that the second term in the asymptotic expansion of the analytical solution is of the 6th order. One can consult [43, 73, 111] for all the details. McTigue solution, from a practical point of view, does not help to identify the pressure and volume separately [127].

Thanks to their simplicity, the Mogi and Mctigue models allow easily and quickly to compare their solutions with real deformation data in terms of physical parameters related to the cavity, i.e., depth and the product volume times pressure. These are the reasons for which these models have been and are widely applied in the forecasting of volcanic eruptions [73].

For completeness of information, we mention that the Mogi model is also called *point dilatation model* sometimes. This is due to the fact that since the ratio $\varepsilon := r_0/d_0$ is small, the magma chamber is well-approximated by a single point producing a uniform pressure in the radial direction. We notice that the asymptotic expansion of the solution of the Mogi model still contains the memory of the spherical shape of the cavity. Therefore if we take different geometrical shapes we get different deformation effects [35, 36, 51, 73]. This aspect will be clearer in Chap. 3, once an asymptotic expansion for a cavity of generic shape will be provided.

With no claim of completeness, we conclude this short introduction providing briefly a summary of other models which have been proposed in the last few decades to overcome some of the limitations and simplifications of the Mogi and McTigue models.

Models based on ellipsoidal cavities or dislocations have been considered in order to explain deviation of the displacements field (measurements) from radial symmetries. In fact, a more accurate modeling of a volcanic system has also to consider the presence of a conduit through which the magma reaches the surface. Bonaccorso and Davis [51] formulated a model based on a prolate spheroid to describe surface deformation due to the presence of a closed conduit, and a dislocation one for magma rising in a conduit that is open at the top. Moreover, we mention solutions on prolate ellipsoids provided by Yang et al. in [137] and on sill-like magma intrusions or chambers which are represented by a finite rectangular dislocations [67, 121], oblate ellipsoids [68], or horizontal cracks [82].

One limitation of the Mogi model (and consequently of the McTigue model) is the idealization of the Earth as a flat half-space. In fact, in [64, 135] the authors consider a non-flat half-space in order to describe the topography of the Earth's crust. In addition, other limitations of the models above are the description of the Earth crust as a homogeneous medium and the uniform pressure on the boundary of the cavity. From the structural point of view, the crust has stratifications with different mechanical properties which influence ground's deformations. Moreover, due to the high temperatures, the solid crust adjacent to a magma chamber cannot be described by a purely elastic medium. Therefore, models with layers with a nonuniform pressure distribution acting on the boundary of the cavity and with a viscoelastic manner around the magma chamber could be more appropriate. See discussion in [69, 92, 94, 108, 127, 129].

We finally mention the attempts of combining elastic properties with gravitational effects through elasto-dynamic equations [80, 81] or viscoelastic rheologies (after [41] and [65]).

All these last models share the idea to describe the physical reality as refined as possible. This is an important target but they require a detailed knowledge of the mechanical properties of Earth's crust which are sometimes missing. This is the reason for which Mogi, McTigue, and other models described above, such as the case of prolate and oblate cavities, continue to be applied in geological literature.

In the following section we summarize the principal aim of this book, showing the mathematical generalization of the Mogi model to the case of a cavity with an arbitrary shape.

1.1 The Mathematical Model

In this section we introduce the boundary value problem behind the Mogi model. We first recall the definition of $C^{k,\alpha}$ regularity for a bounded domain.

Definition 1.1.1 ($C^{k,\alpha}$ **Domain Regularity**) Let Ω be a bounded domain in $\mathbb{R}^d$. Given k, α, with $k \in \mathbb{N}$ and $0 < \alpha \leq 1$, we say that a portion Σ of $\partial\Omega$ is of class $C^{k,\alpha}$ with constant r_0, L_0, if for any $\boldsymbol{P} \in S$, there exists a rigid transformation of coordinates under which we have $\boldsymbol{P} = \boldsymbol{0}$ and

$$\Omega \cap B_{r_0}(\boldsymbol{0}) = \{\boldsymbol{x} \in B_{r_0}(\boldsymbol{0}) \,:\, x_3 > \psi(\boldsymbol{x}')\},$$

where ψ is a $C^{k,\alpha}$ function on $B'_{r_0}(\boldsymbol{0}) \subset \mathbb{R}^{d-1}$ such that

$$\begin{aligned} \psi(\boldsymbol{0}) &= 0, \\ \nabla\psi(\boldsymbol{0}) &= \boldsymbol{0}, \qquad \text{for } k \geq 1 \\ \|\psi\|_{C^{k,\alpha}(B_{r_0}(\boldsymbol{0}))} &\leq L_0. \end{aligned}$$

We say that Σ is of Lipschitz class with constants r_0, L_0 in the case $k = 0, \alpha = 1$.

Most of the results of this book are valid for Lipschitz domains. We will always indicate when more regularity on the domain is needed.

One of our aims is to generalize Mogi model to the case of a magma chamber, i.e., the cavity, of a generic shape. We denote by

$$\mathbb{R}^3_- = \{\boldsymbol{x} \in \mathbb{R}^3,\ x_3 < 0\}$$

the half-space and by $C \subset \mathbb{R}^3_-$ an open set, which has a connected and bounded Lipschitz boundary, describing the magma chamber. In this way, Earth's crust is represented by $\mathbb{R}^3_- \setminus \overline{C}$. By $\mathbb{R}^2 := \{\boldsymbol{y} = (y_1, y_2, y_3) \in \mathbb{R}^3 : y_3 = 0\}$, we denote the free air/crust border, and by ∂C the corresponding crust/chamber border. We consider a homogeneous and isotropic medium subjected to small elastic deformations. Given a matrix $\mathbf{A} \in \mathbb{R}^{3\times 3}$, we denote by $\widehat{\mathbf{A}}$ its symmetric part, that is $\widehat{\mathbf{A}} = \frac{1}{2}(\mathbf{A} + \mathbf{A}^T)$. We study the following elastostatic problem:

$$\begin{cases} \operatorname{div}(\mathbb{C}\widehat{\nabla}\boldsymbol{u}) = \mathbf{0} & \text{in } \mathbb{R}^3_- \setminus C \\ (\mathbb{C}\widehat{\nabla}\boldsymbol{u})\boldsymbol{n} = p\,\boldsymbol{n} & \text{on } \partial C \\ (\mathbb{C}\widehat{\nabla}\boldsymbol{u})\boldsymbol{e}_3 = \mathbf{0} & \text{on } \mathbb{R}^2 \\ \boldsymbol{u} = o(\mathbf{1}), \quad \nabla\boldsymbol{u} = o(|\boldsymbol{x}|^{-1}) & |\boldsymbol{x}| \to \infty, \end{cases} \tag{1.2}$$

where $\boldsymbol{u}$ is the displacement field; $\mathbb{C} := \lambda\mathbf{I} \otimes \mathbf{I} + 2\mu\mathbb{I}$ is the fourth-order isotropic elasticity tensor with $\mathbf{I}$ the identity matrix and $\mathbb{I}$ the fourth-order tensor defined by $\mathbb{I}\mathbf{A} := \widehat{\mathbf{A}}$; p is a constant representing the pressure.

In this book we primarily collect some of the results in [35, 37], regarding the well-posedness of problem (1.2) which can be obtained in two different ways: one way is to prescribe the decay conditions at infinity for $\boldsymbol{u}$, as in (1.2), and then to provide an integral representation formula for $\boldsymbol{u}$ by means of single and double layer potentials of linear elasticity. Then, the well-posedness derives from the study of some integral equations. Another way is to set the analysis of (1.2) in suitable weighted Sobolev spaces where the decay conditions are expressed by weights. In this way we do not have to prescribe explicit decay conditions in (1.2). In the last decades the approach of weighted Sobolev spaces has been deeply investigated by Amrouche and collaborators for solving many mathematical problems in unbounded domains. For more details, see, for example, [28–33] and the references therein. For a related approach, see also Volpert's monograph [133].

Once the well-posedness of the forward problem has been provided, we mainly present and study two different problems:

Problem A: Analyze uniqueness and find stability estimates for the inverse problem of determining the pressurized cavity C from measurements on the boundary of the half-space.

Problem B: Deduce a corresponding point source model, in the same spirit as the Mogi spherical model.

The two different ways of studying the well-posedness of the direct problem are employed to solve Problem A and B. In particular, the first way (based on an integral approach) is useful to solve Problem B. We take inspiration from the works of Ammari and Kang [21, 27]. The second way (based on weighted functional spaces) is utilized to get quantitative stability estimates of the weak solution of (1.2), which are then employed for the analysis of Problem A. Here we follow the approach of

Morassi and Rosset [115, 116]. In the following we give some ideas of the two problems introduced above.

The inverse problem (Problem A) can be stated precisely in this way: is it possible to detect uniquely and in a stable way the cavity C from partial measurements on $\mathbb{R}^2$ of the displacement field $\boldsymbol{u}$? To solve this problem, we assume that the pressure p in (1.2) is known. We briefly explain what we mean by stability estimates for the inverse problem that we are treating. We presume to have data in $B'_{s_0}(\mathbf{0}) = \{\boldsymbol{x} \in \mathbb{R}^2 : x_1^2 + x_2^2 < s_0^2\}$, where $s_0 > 0$, which is a subset of $\mathbb{R}^2$. Moreover, let us assume that there exist two domains C_1 and C_2, satisfying some suitable regularity assumptions (which will be specified in Chap. 3, Sect. 3.3), and $\boldsymbol{u}_1, \boldsymbol{u}_2$ solutions to (1.2) with $C = C_i$ for $i = 1, 2$. If

$$\|\boldsymbol{u}_1 - \boldsymbol{u}_2\|_{L^2(B'_{s_0}(\mathbf{0}))} \leq \epsilon,$$

where $\epsilon > 0$, then we want to prove that

$$d_{\mathcal{H}}(\partial C_1, \partial C_2) \leq c f(\epsilon), \tag{1.3}$$

where $d_{\mathcal{H}}(\partial C_1, \partial C_2)$ is the Hausdorff distance between the two sets $\partial C_1, \partial C_2$, c is a quantitative constant and f is a function that goes to zero as ϵ goes to zero.

In Problem B, we assume that the cavity C is of the form

$$C = d_0 z + r_0 \Omega,$$

where $d_0, r_0 > 0$. In particular, $d_0 z \in \mathbb{R}^3_-$ and Ω (a bounded Lipschitz domain containing the origin) are the center and the shape of the cavity, respectively. By assuming that $d_0 >> r_0$, hence rescaling variables of (1.2) in an appropriate manner, i.e., $(\boldsymbol{x}, \boldsymbol{u}) \mapsto (\boldsymbol{x}/d_0, \boldsymbol{u}/r_0)$ and, denoting the new variables again by $\boldsymbol{x}$ and $\boldsymbol{u}$, we rewrite (1.2) in the form

$$\begin{cases} \operatorname{div}(\mathbb{C}\widehat{\nabla}\boldsymbol{u}) = \mathbf{0} & \text{in } \mathbb{R}^3_- \setminus \overline{C}_\varepsilon \\ (\mathbb{C}\widehat{\nabla}\boldsymbol{u})\boldsymbol{n} = p\,\boldsymbol{n} & \text{on } \partial C_\varepsilon \\ (\mathbb{C}\widehat{\nabla}\boldsymbol{u})\boldsymbol{e}_3 = \mathbf{0} & \text{in } \mathbb{R}^2 \\ \boldsymbol{u} = o(\mathbf{1}), \quad \nabla\boldsymbol{u} = o(|\boldsymbol{x}|^{-1}) & |\boldsymbol{x}| \to \infty, \end{cases} \tag{1.4}$$

where $\varepsilon = r_0/d_0$, $C_\varepsilon := z + \varepsilon\Omega$, and p is a "rescaled" pressure, ratio of the original pressure p and ε. Following the approach of point source model (1.1), we determine an asymptotic expansion, valid for $\boldsymbol{y} \in \mathbb{R}^2$, of the form

$$\boldsymbol{u}_\varepsilon(\boldsymbol{y}) = \varepsilon^\alpha p\, \boldsymbol{U}(z, \boldsymbol{y}) + o(\varepsilon^\alpha) \qquad \text{as } \varepsilon \to 0^+$$

for some appropriate exponent $\alpha > 0$ and principal term $\boldsymbol{U}$. We mention that the explicit solution of Mogi's model, i.e., (1.1), corresponds to the asymptotic expansion of solution of (1.4) with $C_\varepsilon = z + \varepsilon B_1(\mathbf{0})$, that is, $\Omega = B_1(\mathbf{0})$ is the ball of center $\mathbf{0}$ and radius 1.

Before going into details in studying of the previous problems, we provide, in next section, an overview of the mathematical literature on these topics.

1.2 A Brief Overview of the Mathematical Literature

One obstacle in studying inverse problems is that the solution is very sensitive to variations in data. Otherwise stated, the solution of the inverse problem does not depend on the measured data in a Lipschitz continuous way. To clarify this point we need a notion of well-posedness. The concept of well-posedness of a mathematical problem is usually attributed to Hadamard [89, 90]. He proposed that a well-posed problem needs to have three properties: it exists a solution (existence), that is unique (uniqueness) and that depends continuously on the data (stability). We say that a problem is ill-posed if one of these requisites is not fulfilled. Calderón and Zygmund [55, 57, 58] and John [96, 97], see also the references therein, pointed out that many mathematical problems, arising from applications, are ill-posed. For instance, the inverse boundary value problem considered in this book, on detection and reconstruction of unknown cavities from boundary measurements, is non-linear and severely ill-posed. This consideration comes from the important results obtained by Morassi and Rosset in [115] and [116] where, in the context of the isotropic linear elasticity in bounded domains, the problems of detecting cavities and rigid inclusion, respectively, from a single pair of Cauchy data have been investigated. In particular, in these papers the authors proved that with some a priori assumptions on the regularity of the domain to be detected ($C^{1,\alpha}$ regularity, with $0 < \alpha \leq 1$), the continuous dependence (stability) of the domain from the measured data is of log-log type (i.e., the function f in (1.3) has the form $1/(\ln|\ln \epsilon|)^\eta$, for a suitable $0 < \eta \leq 1$). Their outcomes are based on some previous results due to [7, 46] on detection of cavities or of an inaccessible part of the boundary of the domain, in the framework of the scalar elliptic equation (electric conductors). The authors were able to prove stability estimates of log type. This kind of estimates turn out to be the best possible as proved in Alessandrini [3] and Alessandrini-Rondi [5] in space dimension 2, independently from the number of measurements. Indeed the authors showed through explicit examples that assuming up to any infinite order of differentiability of the boundary of the domain to be detected, the continuous dependence (stability) of the domain from the measured data is at the best logarithmic. An inverse problem is said to be exponentially ill-posed, or severely ill-posed, if the optimal stability estimates are at most of logarithmic type. For scalar elliptic equations and stability estimates for the Cauchy problem we refer to the fundamental review paper [14].

The main tools utilized both in [7] (scalar case) and in [115] (elastic case) in determining stability estimates consist in suitable quantitative estimates of unique continuation in the interior and at the boundary. The main difference between the two works is that for solutions to the Lamé system a doubling inequality at the boundary is missing. This is the reason for which in [115] only log-log estimates have been found. However, to overcome the lack of a doubling inequality at the boundary, the authors in [115] had to prove a refined version of a stability estimate of continuation from the interior, which had proved to be extremely useful in the treatment of various inverse boundary value problems associated with a scalar elliptic equation, see [14].

The list of the results introduced above is only partial. It is difficult and impossible to list all the outcomes in this line of research and related topics. With no ambition of completeness, for recent results on detection of rigid inclusions see [15, 117], for inverse problems for thin elastic plates see [16, 118–120]. For detection of cracks in the electrostatic case see [6], in the isotropic and homogeneous elastic case [49]. See the references therein.

We highlight that the ill-posedness constitutes a severe difficulty for the numerical treatment of inverse problems. This means that without adding other information on the problem, reconstruction algorithms cannot provide good quality results. On the other hand, if we add more structural information about the medium and/or the problem, we may restore the well-posedness, i.e., Lipschitz continuous dependence from boundary measurements. For instance, as pointed out by Friedman and Vogelius in [85], for the problem of detecting inhomogeneities inside a domain, one can retrieve the well-posedness of the inverse problem if well separated inhomogeneities of small size are considered. In [85] the authors analyzed the electrostatic problem for a conductor with many small regular inhomogeneities of extreme conductivities (infinite or zero conductivity) providing the leading term of the asymptotic formula for the perturbed potential. Then, they were able to prove that locations and relative sizes of the inhomogeneities depend Lipschitz-continuously on the potential measurement on any open subset of the boundary. This paper can be considered the pioneering work on the asymptotic expansions in inverse problems. Subsequently, many efforts have been made to improve and generalize these results. For instance, in [61] the authors derived the leading order term of an asymptotic formula for the steady-state voltage potential for a conductor with a finite number of small inhomogeneities with (finite) conductivity different from the background conductivity. In addition, they have also provided a reconstruction procedure based on the asymptotic formula employing a fairly straightforward least-square approach (see [61] for a more in-depth discussion). For other algorithms that makes use of an asymptotic expansion of the voltage potentials see for instance [25, 54]. The case of a finite set of inhomogeneities of small thickness, in the two-dimensional space, has been considered in [47]. In [48, 59] the authors, in a most natural way, generalized and unified the specific formulas already derived for a finite set of inhomogeneities of small diameter, see [61], and for a finite set of inhomogeneities of small thickness, see [47], to the case of inhomogeneities of low volume fraction. See also [60] for the case of multiple boundary measurements.

Analogously to [61], the case of the time-harmonic Maxwell's equations with "transverse electric" and "transverse magnetic" symmetries has been studied in

[132]. The authors provided the leading term of an asymptotic formulas for the electric and magnetic fields on the boundary of the domain due to the presence of small objects with different material characteristics (supposed constants) with respect to the medium. These results have been generalized to the time-harmonic Maxwell's equation in [23].

Isotropic elastic media were considered in [17]. The authors found an asymptotic formula for the displacement vector field in terms of the reference Lamé coefficients, the location of the imperfections, and their geometry. Then, they used this asymptotic expansion to establish continuous dependence estimates for certain characteristics of the imperfections in terms of the boundary data and to derive integral boundary formulae for a reconstruction procedure. For thin elastic inhomogeneities and crack see [45] and [49], respectively.

Using a different mathematical approach, based on layer potential techniques, Ammari and Kang were able to derive higher-order terms in the asymptotic expansions, of most problems above, in the more general framework of Lipschitz domains. For the results on the conductivity equation see [19], on the Helmholtz equation [20] and on the Lamé system [24]. All the works above share the philosophy that by means of partial or complete asymptotic formulas of solutions to boundary value problems and some efficient algorithms, information about the location and size of the inclusions can be reconstructed, see [21, 27, 88].

The literature of the two last decades on this field is really vast so we cannot cover all the results. For this reason, we refer the reader to [18, 21, 22, 27] and the works therein for a more extensive literature.

The approach introduced by Ammari and Kang proved to be a very powerful and elegant method to obtain asymptotic expansion of any order for solutions to the transmission problems and, as a particular case, to cavities and perfectly conducting inclusions' problem. Using this approach, in [35, 36], the authors extended the analysis to the case of a pressurized cavity, i.e., a hole with nonzero tractions on its boundary, in a half-space.

Guided by the historical approach summarized above for which the electrostatic problem was the first one considered in the field of the asymptotic analysis, in Chap. 2 we analyze a simplified scalar version of the elastic model introduced in Sect. 1.1. This toy model is based on harmonic functions in a half-space. As far as we know it does not have a physical application but the mathematical result has an interest on its own. Most of the results contained in this chapter come from [35]. In Chap. 3, we analyze the elastic model of Sect. 1.1 and in particular we give solutions to Problem A and B. The results of this chapter are essentially contained in [36] and [37].

Chapter 2
A Scalar Model in the Half-Space

The aim of this chapter is to provide a detailed mathematical study of a simplified scalar version of the elastic problem presented in Chap. 1. Since this problem is easier than the elastic case it will be studied in the generic framework of a d-dimensional space, with $d \geq 3$. In particular, we indicate with $\mathbb{R}^d_-$ the d-dimensional half-space and with $\mathbb{R}^{d-1}$ its boundary. Then we consider the following boundary value problem:

$$\begin{cases} \Delta u = 0 & \text{in } \mathbb{R}^d_- \setminus C \\ \dfrac{\partial u}{\partial \boldsymbol{n}} = g & \text{on } \partial C \\ \dfrac{\partial u}{\partial x_d} = 0 & \text{on } \mathbb{R}^{d-1} \\ u(\boldsymbol{x}) \to 0 & \text{as } |\boldsymbol{x}| \to +\infty, \end{cases} \tag{2.1}$$

where C is a cavity (analogous to the pressurized one for the elastic case), g is a function defined on ∂C. After proving the well-posedness of this boundary value problem, we will consider the case of a cavity of the form $C = z + \varepsilon\Omega$, where $z \in \mathbb{R}^d_-$ and Ω is a Lipschitz bounded domain containing the origin. Our aim is to establish an asymptotic expansion of the solution of the problem as $\varepsilon \to 0$ in the same spirit as the point dilatation model presented in the previous chapter.

As far as we know, this model does not have a real physical application; however, the mathematical result has an interest on its own. In fact, as explained in Chap. 1, it belongs to the same stream of the asymptotic analysis of the conductivity equation in bounded domains.

The results we are going to present come from [35]. The main novelty concerns the analysis of the asymptotic expansion of a pressurized cavity buried in an unbounded domain. This kind of theme was not covered by the mathematical literature, despite the presence of a lot of works on asymptotic expansion techniques [18, 21, 22, 25, 26, 47, 48, 59, 85].

A. Aspri, *An Elastic Model for Volcanology*, Lecture Notes in Geosystems Mathematics and Computing, https://doi.org/10.1007/978-3-030-31475-0_2

The main mathematical tools that we apply for the analysis of this problem are the single and double layer potentials, which result in studying integral equations. This approach is not new but it has been proposed by Ammari and Kang, see, for instance, [21, 22, 27].

This chapter is organized as follows: In Sect. 2.1 we provide a short presentation of the classic Sobolev spaces, weighted Sobolev spaces, and properties of single and double layer potentials for harmonic functions. Readers who are familiar with topics can skip this part and go directly to Sect. 2.3 where the boundary value problem (2.1) is analyzed. We provide its well-posedness both using integral equations and a variational approach based on weighted Sobolev spaces. In Sect. 2.4, we state and prove a spectral result on the integral potentials involved in the integral representation formula of the solution of (2.1). This result can be used for the derivation of the asymptotic expansion when the cavity C is more regular than Lipschitz. In the last Sect. 2.5 we provide the asymptotic expansion of the solution of (2.1).

2.1 Sobolev Spaces: Classic and Weighted

Sobolev spaces are the proper setting where to study partial differential equations. In this section we give a brief survey of the main results on this topic. In the first part we recall the classic Sobolev spaces defined in a bounded domain while in the second part we present weighted Sobolev spaces in the setting of a half-space. For more in-depth discussion on these topics we propose the reading of [30, 76, 101].

Given Ω a bounded domain, we first recall the definition of L^p Banach spaces.

Definition 2.1.1 Let $p \in \mathbb{R}$ such that $p \geq 1$. We define

$$L^p(\Omega) = \left\{\text{“class of functions” } \varphi : \Omega \to \mathbb{R} \text{ s. t. } \int_\Omega |\varphi(\boldsymbol{x})|^p \, d\boldsymbol{x} < +\infty\right\},$$

with the norm given by

$$\|\varphi\|_{L^p(\Omega)} = \left(\int_\Omega |\varphi(\boldsymbol{x})|^p \, d\boldsymbol{x}\right)^{1/p},$$

and

$$L^\infty(\Omega) = \left\{\text{“class of functions” } \varphi : \Omega \to \mathbb{R} : \ \exists c \text{ s.t. } |\varphi(\boldsymbol{x})| \leq c, \ \text{ a.e. } \boldsymbol{x} \in \Omega\right\},$$

with the norm given by

$$\|\varphi\|_{L^\infty(\Omega)} = \inf\left\{c \text{ s.t. } |\varphi(\boldsymbol{x})| \leq c \text{ a.e. } \boldsymbol{x} \in \Omega\right\}.$$

In addition we denote by $L^p_{\text{loc}}(\Omega)$ the set of functions ψ which are defined in Ω and such that $\psi \in L^p(D)$ for any bounded domain $D \subset\subset \Omega$, i.e., for any bounded domain D such that $\overline{D} \subset \Omega$.

The first step to define Sobolev spaces is to weaken the notion of derivative. To this aim, we define

$$\mathcal{D}(\Omega) = C_0^\infty(\Omega), \tag{2.2}$$

that is, the space of infinitely differentiable functions $\varphi : \Omega \to \mathbb{R}$, with compact support in Ω and $\mathcal{D}'(\Omega)$ the dual space of $\mathcal{D}(\Omega)$. Sometimes we will refer to this space as the space of *test functions*. Using the multi-index notation, $\alpha = (\alpha_1, \cdots, \alpha_d)$, with $\alpha_i \geq 0$, integers, we define $D^\alpha \varphi$ as

$$D^\alpha = \frac{\partial^{|\alpha|}}{\partial x_1^{\alpha_1} \cdots \partial x_d^{\alpha_d}}, \qquad d \geq 1.$$

In the next, we use the most common notation $\nabla \cdot$ to represent the gradient of a function.

Definition 2.1.2 Assume that $v, w \in L^1_{\text{loc}}(\Omega)$, and α is a multi-index. We say that w is the α-weak derivative of v if

$$\int_\Omega v(\boldsymbol{x}) D^\alpha \varphi(\boldsymbol{x})\, d\boldsymbol{x} = (-1)^{|\alpha|} \int_\Omega w(\boldsymbol{x}) \varphi(\boldsymbol{x})\, d\boldsymbol{x}.$$

With this definition at hand, we can define the classic Sobolev spaces

Definition 2.1.3 Let $m > 0$ be an integer and $1 \leq p \leq +\infty$. The Sobolev space $W^{m,p}(\Omega)$ is defined as the space of function $v \in L^p(\Omega)$ such that for every multi-index α with $|\alpha| \leq m$, the weak derivative $D^\alpha v$ exists and $D^\alpha v \in L^p(\Omega)$, i.e.,

$$W^{m,p}(\Omega) = \Big\{ v \in L^p(\Omega) : \ |\alpha| \leq m, \ D^\alpha v \in L^p(\Omega) \Big\}.$$

These function spaces are Banach spaces equipped with their natural norm

$$\|v\|^p_{W^{m,p}(\Omega)} = \sum_{|\alpha|=0}^{m} \|D^\alpha v\|^p_{L^p(\Omega)}, \qquad \text{for } 1 \leq p < \infty$$

$$\|v\|_{W^{m,\infty}(\Omega)} = \sum_{|\alpha|=0}^{m} \|D^\alpha v\|_{L^\infty(\Omega)}.$$

Moreover, we denote by $W_0^{m,p}(\Omega)$ the space

$$W_0^{m,p}(\Omega) = \overline{\mathcal{D}(\Omega)}^{\|\cdot\|_{W^{m,p}(\Omega)}}.$$

In the particular case where $p = 2$, that is, in the Hilbert setting which we consider in this book, we use the following notation:

$$H^m(\Omega) = W^{m,2}(\Omega) \qquad \text{and} \qquad H_0^m(\Omega) = W_0^{m,2}(\Omega), \quad \text{for } k \geq 0.$$

We conclude recalling the trace theorem which defines the traces of functions in $H^m(\Omega)$ on $\partial\Omega$, as its name indicates.

Theorem 2.1.1 (Theorem 2.7.4 [101]) *Let Ω be a bounded domain of class C^{m+1}. Then there exists a trace map $\gamma = (\gamma_0, \gamma_1, \cdots, \gamma_{m-1})$ from $H^m(\Omega)$ into $L^2(\partial\Omega)^m$ such that*

(i) if $v \in C^\infty(\overline{\Omega})$ then

$$\gamma_0(v) = v_{|\partial\Omega}, \quad \gamma_1(v) = \frac{\partial v}{\partial \boldsymbol{n}}\Big|_{\partial\Omega} \quad \cdots \quad , \gamma_{m-1}(v) = \frac{\partial^{m-1} v}{\partial \boldsymbol{n}^{m-1}}\Big|_{\partial\Omega}$$

where $\boldsymbol{n}$ is the outward unit normal vector.

(ii) The range of γ is

$$\prod_{k=0}^{m-1} H^{m-k-\frac{1}{2}}(\partial\Omega).$$

(iii) The kernel of γ is $H_0^m(\Omega)$.

2.1.1 Weighted Sobolev Spaces

We recall here the definition of weighted Sobolev spaces, restricting our attention to $\mathbb{R}^d_- \setminus \Omega$ (for $d = 3$ we have the unbounded set used in this book), and we deal only with L^2-based spaces to avoid unnecessary technicalities. Moreover, we mention that we do not consider the case of logarithmic weights so to maintain the presentation of the topics as readable as possible and since for our purposes are not necessary. For the readers interested in general results on weighted Sobolev spaces with polynomial and logarithmic weights we suggest to consult [28–33] and the references therein. Again, we denote the space of indefinitely differentiable functions with compact support in $\mathbb{R}^d_- \setminus \overline{\Omega}$ by $\mathcal{D}(\mathbb{R}^d_- \setminus \overline{\Omega})$ and its dual space by $\mathcal{D}'(\mathbb{R}^d_- \setminus \overline{\Omega})$, i.e., the space of distributions.

Definition 2.1.4 (Weighted Sobolev Space) Let

$$\rho = (1 + |\boldsymbol{x}|^2)^{1/2}. \tag{2.3}$$

For $m \in \mathbb{Z}_+$, $\iota \in \mathbb{R}$, we define

$$H_\iota^m(\mathbb{R}^d_- \setminus \overline{\Omega}) = \left\{ v \in \mathcal{D}'(\mathbb{R}^d_- \setminus \overline{\Omega}); 0 \leq |\kappa| \leq m, \rho^{\iota - m + |\kappa|} D^\kappa v \in L^2(\mathbb{R}^d_- \setminus \overline{\Omega}) \right\}.$$

These functional spaces are Banach spaces equipped with their natural norm

$$\|v\|^2_{H_\iota^m(\mathbb{R}^d_- \setminus \overline{\Omega})} = \sum_{|\kappa|=0}^{m} \|\rho^{\iota - m + |\kappa|} D^\kappa f\|^2_{L^2(\mathbb{R}^d_- \setminus \overline{\Omega})}.$$

Remark 2.1.2 We recall that the weights are chosen so that the space $\mathcal{D}(\overline{\mathbb{R}^d_-} \setminus \Omega)$ is dense in $H_\iota^m(\mathbb{R}^d_- \setminus \overline{\Omega})$, see [28, 91].

Remark 2.1.3 If we consider $D \subset (\mathbb{R}^d_- \setminus \overline{\Omega})$ then the spaces $H_\iota^m(D)$ reduce to the usual Sobolev spaces $H^m(D)$. Hence the usual trace theorems hold.

Remark 2.1.4 To avoid confusion between the classic Sobolev spaces H_0^m and the weighted Sobolev spaces H_ι^m with $\iota = 0$, we use for the latter the notation H_w^m where w stands for the word "weight." Therefore

$$H_w^m(\mathbb{R}^d \setminus \overline{\Omega}) = \left\{ v \in \mathcal{D}'(\mathbb{R}^d_- \setminus \overline{\Omega}); 0 \leq |\kappa| \leq m, \rho^{-m + |\kappa|} D^\kappa v \in L^2(\mathbb{R}^d_- \setminus \overline{\Omega}) \right\}.$$

In addition, we define

$$\mathring{H}_\iota^m(\mathbb{R}^d_- \setminus \overline{\Omega}) = \overline{\mathcal{D}(\mathbb{R}^d_- \setminus \overline{\Omega})}^{\|\cdot\|_{H_\iota^m(\mathbb{R}^d_- \setminus \overline{\Omega})}}, \tag{2.4}$$

which is a proper subset of $H_\iota^m(\mathbb{R}^d_- \setminus \overline{\Omega})$.

Then, we define $H_{-\iota}^{-m}(\mathbb{R}^d_- \setminus \overline{\Omega})$ as the dual space to $\mathring{H}_\iota^m(\mathbb{R}^d_- \setminus \overline{\Omega})$. Both are spaces of distributions.

Before defining fractional spaces and stating the trace theorem, we recall some basic properties of the spaces H_ι^m. For simplicity, we state most of these properties, such as Poincaré and Korn inequality only for the space $H_w^1(\mathbb{R}^3_- \setminus \overline{\Omega})$, which is the only weighted Sobolev space utilized in this book. The following weighted Poincaré-type inequality holds (see for example [29, 91]):

$$\|f\|_{H_w^1(\mathbb{R}^3_- \setminus \overline{\Omega})} \leq C_1 \|\nabla f\|_{L^2(\mathbb{R}^3_- \setminus \overline{\Omega})}. \tag{2.5}$$

Similarly, if a vector field $\boldsymbol{u}$ has square-integrable deformation tensor $\widehat{\nabla} \boldsymbol{u}$ and belongs to the space $H_{-1}^0(\mathbb{R}^3_- \setminus \overline{\Omega})$, then $\boldsymbol{u} \in H_w^1(\mathbb{R}^3_- \setminus \overline{\Omega})$ and a Korn-type inequality holds, see [32] and [102], i.e., there exists a positive constant C such that

$$\|\nabla \boldsymbol{u}\|_{L^2(\mathbb{R}^3_- \setminus \overline{\Omega})} \leq C \, \|\widehat{\nabla} \boldsymbol{u}\|_{L^2(\mathbb{R}^3_- \setminus \overline{\Omega})}. \tag{2.6}$$

In Chap. 3 we will provide a quantitative version of these two inequalities in $\mathbb{R}^3 \setminus \overline{\Omega}$ useful to solve the inverse problem presented in Chap. 1.

To define the trace operator we need to introduce the fractional spaces on $\mathbb{R}^d$. For the trace on $\partial\Omega$, see the trace theorem for bounded domain introduced in the previous section.

We only deal with the case $d = 2, 3$ and, in particular, we give the definition in $\mathbb{R}^3$ for $0 < s < 2$. The same definition applies in $\mathbb{R}^2$ for $0 < s < 1$. In both cases, we assume $\iota \geq -1$ and $d/2 + \iota \neq s$, $d = 2, 3$.

For $0 < s < 1$, we define the weighted fractional space $H^s_\iota(\mathbb{R}^3)$ as

$$H^s_\iota(\mathbb{R}^3) = \left\{ v \in \mathcal{D}'(\mathbb{R}^3);\ \varrho^{\iota-s}\, v \in L^2(\mathbb{R}^3),\ |v|_s < \infty \right\}, \tag{2.7}$$

where $|\cdot|_s$ denotes a weighted seminorm:

$$|v|_s := \iint\limits_{\mathbb{R}^3\times\mathbb{R}^3} \frac{|\varrho^\iota(\boldsymbol{x})\, v(\boldsymbol{x}) - \varrho^\iota(\boldsymbol{y})\, v(\boldsymbol{y})|^2}{|\boldsymbol{x}-\boldsymbol{y}|^{3+2s}}\, d\boldsymbol{x}\, d\boldsymbol{y}.$$

For $1 < s < 2$, we define $H^s_\iota(\mathbb{R}^3)$ as

$$H^s_\iota(\mathbb{R}^3) = \left\{ v \in \mathcal{D}'(\mathbb{R}^3);\ 0 \leq |\kappa| \leq [s]-1,\ \varrho^{\iota-s+|\kappa|}\, D^\kappa v \in L^2(\mathbb{R}^3), \right. \\ \left. D^{[s]} v \in H^{s-[s]}_\iota(\mathbb{R}^3) \right\}, \tag{2.8}$$

with $[s]$ the greatest integer less than or equal to s. From this space, we can immediately define by restriction the fractional spaces on $\mathbb{R}^3_-$, i.e.,

$$H^s_\iota(\mathbb{R}^3_-) := \left\{ v \in \mathcal{D}'(\overline{\mathbb{R}^3_-}) : \exists\, w \in H^s_\iota(\mathbb{R}^3),\ v = w_{|\mathbb{R}^3_-} \right\},$$

equipped with the norm

$$\|v\|_{H^s_\iota(\mathbb{R}^3_-)} := \inf\{\|w\|_{H^s_\iota};\ w \in H^s_\iota(\mathbb{R}^3),\ v = w_{|\mathbb{R}^3_-}\}.$$

We conclude this section with the trace theorem for weighted Sobolev spaces.

Proposition 2.1.5 ([32], Lemma 1.1) *Let $m \geq 1$ and $\iota \in \mathbb{R}$. Then, there exists a continuous linear mapping*

$$\boldsymbol{\gamma} = (\gamma_0, \cdots, \gamma_{m-1}) : H^m_\iota(\mathbb{R}^3_-) \longrightarrow \prod_{j=0}^{m-1} H^{m-j-1/2}_\iota(\mathbb{R}^2). \tag{2.9}$$

In addition, $\boldsymbol{\gamma}$ is surjective and

$$\operatorname{Ker} \boldsymbol{\gamma} = \mathring{H}^m_\iota(\mathbb{R}^3_-).$$

2.2 Harmonic Functions and Layer Potentials

We recall some important properties about the decay rate of harmonic functions in unbounded domains and single and double layer potentials for the Laplace operator on Lipschitz domains. As already explained at the beginning of this chapter, we focus the attention only to dimension $d \geq 3$; however, we remark that most of the results recalled in this section are true also for $d = 2$.

We skip the proofs of the basic concepts while we give them for some theorems that may be unfamiliar. Results about harmonic functions in unbounded domains are contained, for instance, in [76, 84, 124]; those on properties of single and double layer potentials can be found in [21, 75, 103, 131].

2.2.1 *Some Decay Properties*

To show the well-posedness of the boundary value problem (2.1) by means of single and double layer potentials, we can use the symmetry of the half-space in order to extend the problem to an exterior domain in the full space. Therefore, we first need to recall some classical results on the behavior of harmonic functions in $\mathbb{R}^d$. These properties can be found for instance in [76, 84, 103].

In the following statements, again we assume that Ω is a bounded domain in $\mathbb{R}^d$, with $d \geq 3$.

Theorem 2.2.1 *If v is harmonic in $\mathbb{R}^d \setminus \Omega$, with $d \geq 3$, the following statements are equivalent.*

1. *v is harmonic at infinity.*
2. *$v(\boldsymbol{x}) \to 0$ as $|\boldsymbol{x}| \to \infty$.*
3. *$|v(\boldsymbol{x})| = O\left(|\boldsymbol{x}|^{2-d}\right)$ as $|\boldsymbol{x}| \to \infty$.*

In addition, from the behavior of the gradient of harmonic functions on the boundary of the *d-dimensional* balls, i.e., if v is a harmonic function in $B_R(\boldsymbol{x})$, it follows that

$$|\nabla v| \leq \frac{d}{R} \max_{\partial B_R(\boldsymbol{x})} |v|, \tag{2.10}$$

which through the Kelvin transform $\tilde{v}(\boldsymbol{x}) = |\boldsymbol{x}|^{2-d} v(|\boldsymbol{x}|^{-2}\boldsymbol{x})$ implies the behavior of the gradient of harmonic functions at infinity. For an in-depth discussion see [84]. Then, we have

Theorem 2.2.2 *If v is harmonic in $\mathbb{R}^d \setminus \Omega$, $d \geq 3$, and $v(\boldsymbol{x}) \to 0$ as $|\boldsymbol{x}| \to \infty$, then there exist r and a constant M, depending on r, such that if $|\boldsymbol{x}| \geq r$, we have*

$$|v| \leq \frac{M}{|\boldsymbol{x}|^{d-2}}, \qquad |\nabla v| \leq \frac{M}{|\boldsymbol{x}|^{d-1}}. \tag{2.11}$$

In conclusion, we recall the Green's second identity which plays a crucial role to convert differential problems into integral ones, i.e., in terms of single and double layer potentials.

Proposition 2.2.3 *Let Ω be a Lipschitz domain in $\mathbb{R}^d$. Given the pair of functions (u, v) defined in Ω it holds*

$$\int_{\Omega} (\Delta u(\boldsymbol{x})v(\boldsymbol{x}) - u(\boldsymbol{x})\Delta v(\boldsymbol{x}))\, d\boldsymbol{x} = \int_{\partial\Omega} \left(\frac{\partial u}{\partial \boldsymbol{n}}(\boldsymbol{x})v(\boldsymbol{x}) - u(\boldsymbol{x})\frac{\partial v}{\partial \boldsymbol{n}}(\boldsymbol{x}) \right) d\sigma(\boldsymbol{x}). \tag{2.12}$$

2.2.2 Single and Double Layer Potentials

Denoting with ω_d the area of the $(d-1)$-dimensional unit sphere, we recall the fundamental solution of the Laplace operator, that is the solution to

$$\Delta \Gamma(\boldsymbol{x}) = \delta_{\boldsymbol{0}}(\boldsymbol{x}),$$

where $\delta_{\boldsymbol{0}}(\boldsymbol{x})$ represents the delta function centered at $\boldsymbol{0}$. It is well known that Γ is radially symmetric and has this expression

$$\Gamma(\boldsymbol{x}) = \frac{1}{\omega_d(2-d)|\boldsymbol{x}|^{d-2}}, \tag{2.13}$$

for $d \geq 3$, see, for example, [76, 84]. Given a bounded Lipschitz domain $\Omega \subset \mathbb{R}^d$, a function $\varphi(\boldsymbol{y}) \in L^2(\partial\Omega)$ and using the fundamental solution (2.13), we introduce the single and double layer potentials (for harmonic functions), i.e., the integral operators

$$\begin{aligned} S_{\Omega}\varphi(\boldsymbol{x}) &:= \int_{\partial\Omega} \Gamma(\boldsymbol{x}-\boldsymbol{y})\varphi(\boldsymbol{y})\, d\sigma(\boldsymbol{y}), && \boldsymbol{x} \in \mathbb{R}^d \\ D_{\Omega}\varphi(\boldsymbol{x}) &:= \int_{\partial\Omega} \frac{\partial \Gamma(\boldsymbol{x}-\boldsymbol{y})}{\partial \boldsymbol{n}_{\boldsymbol{y}}}\varphi(\boldsymbol{y})\, d\sigma(\boldsymbol{y}), && \boldsymbol{x} \in \mathbb{R}^d \setminus \partial\Omega. \end{aligned} \tag{2.14}$$

Here, we summarize some of their properties (for details see [21, 84, 103, 110]) for $d \geq 3$:

(i) By definition, $S_{\Omega}\varphi(\boldsymbol{x})$ and $D_{\Omega}\varphi(\boldsymbol{x})$ are harmonic in $\mathbb{R}^d \setminus \partial\Omega$.
(ii) $S_{\Omega}\varphi(\boldsymbol{x}) = O(|\boldsymbol{x}|^{2-d})$ as $|\boldsymbol{x}| \to +\infty$.
(iii) If $\int_{\partial\Omega} \varphi(\boldsymbol{x})\, d\sigma(\boldsymbol{x}) = 0$ then $S_{\Omega}\varphi(\boldsymbol{x}) = O(|\boldsymbol{x}|^{1-d})$ as $|\boldsymbol{x}| \to +\infty$.
(iv) $D_{\Omega}\varphi(\boldsymbol{x}) = O(|\boldsymbol{x}|^{1-d})$ as $|\boldsymbol{x}| \to +\infty$.

Next, we introduce the Neumann-Poincaré boundary operator $K_{\Omega} : L^2(\partial\Omega) \to L^2(\partial\Omega)$

$$K_\Omega\varphi(\boldsymbol{x}) = \frac{1}{\omega_d} p.v. \int_{\partial\Omega} \frac{(\boldsymbol{y}-\boldsymbol{x})\cdot \boldsymbol{n}_{\boldsymbol{y}}}{|\boldsymbol{x}-\boldsymbol{y}|^d} \varphi(\boldsymbol{y})\, d\sigma(\boldsymbol{y}), \tag{2.15}$$

and its L^2-adjoint

$$K^*_\Omega\varphi(\boldsymbol{x}) = \frac{1}{\omega_d} p.v. \int_{\partial\Omega} \frac{(\boldsymbol{x}-\boldsymbol{y})\cdot \boldsymbol{n}_{\boldsymbol{x}}}{|\boldsymbol{x}-\boldsymbol{y}|^d} \varphi(\boldsymbol{y})\, d\sigma(\boldsymbol{y}), \tag{2.16}$$

where $p.v.$ denotes the Cauchy principal value. The operators K_Ω and K^*_Ω are singular integral operators, bounded on $L^2(\partial\Omega)$.

Given a function v defined in a neighborhood of $\partial\Omega$, we set

$$\begin{aligned} v(\boldsymbol{x})\Big|_\pm &:= \lim_{h\to 0^+} v(\boldsymbol{x} \pm h\boldsymbol{n}_{\boldsymbol{x}}), && \boldsymbol{x} \in \partial\Omega, \\ \frac{\partial v}{\partial \boldsymbol{n}_{\boldsymbol{x}}}(\boldsymbol{x})\Big|_\pm &:= \lim_{h\to 0^+} \nabla v(\boldsymbol{x} \pm h\boldsymbol{n}_{\boldsymbol{x}})\cdot \boldsymbol{n}_{\boldsymbol{x}}, && \boldsymbol{x} \in \partial\Omega. \end{aligned} \tag{2.17}$$

The following theorem about the jump relations of the single and double potentials for Lipschitz domains is a consequence of Coifman-McIntosh-Meyer results on the boundedness of the Cauchy integral on Lipschitz curves, see [63], and the method of rotations of Calderón, see [56].

In the sequel, $t_1, \cdots, t_{d-1}$ represent an orthonormal basis for the tangent plane to $\partial\Omega$ at a point $\boldsymbol{x}$ and $\partial/\partial \boldsymbol{t} = \sum_{k=1}^{d-1} \partial/\partial t_k\, t_k$ the tangential derivative on $\partial\Omega$.

Theorem 2.2.4 *Let $\Omega \subset \mathbb{R}^d$ be a bounded Lipschitz domain. For $\varphi \in L^2(\partial\Omega)$ the following relations hold, a.e., in $\partial\Omega$,*

$$\begin{aligned} S_\Omega\varphi(\boldsymbol{x})\Big|_+ &= S_\Omega\varphi(\boldsymbol{x})\Big|_- \\ \frac{\partial S_\Omega\varphi}{\partial \boldsymbol{t}}(\boldsymbol{x})\Big|_+ &= \frac{\partial S_\Omega\varphi}{\partial \boldsymbol{t}}(\boldsymbol{x})\Big|_- \\ \frac{\partial S_\Omega\varphi}{\partial \boldsymbol{n}_{\boldsymbol{x}}}(\boldsymbol{x})\Big|_\pm &= \left(\pm\frac{1}{2}I + K^*_\Omega\right)\varphi(\boldsymbol{x}) \\ D_\Omega\varphi(\boldsymbol{x})\Big|_\pm &= \left(\mp\frac{1}{2}I + K_\Omega\right)\varphi(\boldsymbol{x}). \end{aligned} \tag{2.18}$$

Using Green's identity it follows that $D_\Omega(1) = 1$; hence, by the jump relations for the double layer potential, we have $K_\Omega(1) = 1/2$.

In the sequel, to determine the well-posedness of the scalar model (2.1), we will need to generalize the result about the invertibility of the operators $1/2I + K^*_\Omega$ and $1/2I + K_\Omega$, when a regular compact operator is added. To this aim, we recall here what is known about the eigenvalues of K^*_Ω and K_Ω in $L^2(\partial\Omega)$ and then the invertibility of the operators $\lambda I - K^*_\Omega$ and $\lambda I - K_\Omega$, for suitable $\lambda \in \mathbb{R}$. These results, for the case of Lipschitz domains, are contained in [75]. We define

$$L_0^2(\partial\Omega) := \Big\{\varphi \in L^2(\partial\Omega), \int_{\partial\Omega} \varphi \, d\sigma = 0\Big\}.$$

Theorem 2.2.5 ([21]) *Let λ be a real number. The operator $\lambda I - K_\Omega^*$ is injective on*

(a) $L_0^2(\partial\Omega)$ if $|\lambda| \geq 1/2$;
(b) $L^2(\partial\Omega)$ if $\lambda \in (-\infty, -\frac{1}{2}] \cup (\frac{1}{2}, +\infty)$.

Proof By contradiction, let $\lambda \in (-\infty, -1/2] \cup (1/2, +\infty)$ and assume there exists $\varphi \in L^2(\partial\Omega)$, not identically zero, satisfying $(\lambda I - K_\Omega^*)\varphi = 0$. Since $K_\Omega(1) = 1/2$, it follows by duality that φ has mean value zero on $\partial\Omega$, in fact

$$\begin{aligned} 0 = \langle 1, (\lambda I - K_\Omega^*)\varphi\rangle_{L^2(\partial\Omega)} &= \langle \lambda - K_\Omega(1), \varphi\rangle_{L^2(\partial\Omega)} \\ &= \langle \lambda - 1/2, \varphi\rangle_{L^2(\partial\Omega)}. \end{aligned}$$

Thus, from the properties of single layer potential, it follows that $S_\Omega\varphi(\boldsymbol{x}) = O(|\boldsymbol{x}|^{1-d})$ and $\nabla S_\Omega\varphi(\boldsymbol{x}) = O(|\boldsymbol{x}|^{-d})$ for $|\boldsymbol{x}| \to \infty$. Since φ is assumed to be not identically zero, the two numbers

$$A = \int_\Omega |\nabla S_\Omega\varphi|^2 \, d\boldsymbol{x}, \qquad B = \int_{\mathbb{R}^d\setminus\overline{\Omega}} |\nabla S_\Omega\varphi|^2 \, d\boldsymbol{x}$$

are not zero. Applying the divergence theorem and the jump relations of the single layer potentials in Theorem 2.2.4 to A and B, we get

$$A = \int_{\partial\Omega} \big(-\frac{1}{2}I + K_\Omega^*\big)\varphi \, S_\Omega\varphi \, d\sigma(\boldsymbol{x}), \qquad B = -\int_{\partial\Omega} \big(\frac{1}{2}I + K_\Omega^*\big)\, \varphi S_\Omega\varphi \, d\sigma(\boldsymbol{x}).$$

Since $(\lambda I - K_\Omega^*)\varphi = 0$, it follows that

$$\lambda = \frac{1}{2}\frac{B - A}{B + A},$$

hence, $|\lambda| < 1/2$, which is a contradiction. This implies that the operator $\lambda I - K_\Omega^*$ is injective in $L^2(\partial\Omega)$ for $\lambda \in (-\infty, -1/2] \cup (1/2, \infty)$.

On the other hand, in the case $\lambda = 1/2$, we suppose by contradiction that there exists $\varphi \in L_0^2(\partial\Omega)$, not identically zero, such that $(1/2I - K_\Omega^*)\varphi = 0$. Then, we define A and B as before, but in this case we find

$$A = \int_{\partial\Omega} \big(-\frac{1}{2}I + K_\Omega^*\big)\varphi \, S_\Omega\varphi \, d\sigma(\boldsymbol{x}) = 0,$$

hence, $S_\Omega\varphi = cost$ in Ω. By the continuity property of single layer potential on $\partial\Omega$ (see Theorem 2.2.4) we have that $S_\Omega\varphi$ is constant on $\partial\Omega$. Moreover, $S_\Omega\varphi$ is harmonic in $\mathbb{R}^d \setminus \partial\Omega$ and behaves like $O(|\boldsymbol{x}|^{1-d})$ as $|\boldsymbol{x}| \to \infty$ because $\varphi \in L^2_0(\partial\Omega)$. Therefore, by the decay rate at infinity we find that $S_\Omega\varphi = 0$ in $\mathbb{R}^d$, hence $\varphi = 0$ on $\partial\Omega$. This contradicts the hypothesis, hence $1/2I - K^*_\Omega$ is injective in $L^2_0(\partial\Omega)$. □

The invertibility results of $\lambda I - K^*_\Omega$ and $\lambda I - K_\Omega$ are not straightforward. If the domain Ω is regular, at least C^1, it can be proven that the boundary operators K_Ω and K^*_Ω are compact; hence, the invertibility of $\lambda I - K^*_\Omega$ and $\lambda I - K_\Omega$ can be obtained by the Fredholm theory. Instead, in the Lipschitz domains, K_Ω and K^*_Ω lose the compactness property, see the example proposed by Fabes, Jodeit, and Lewis in [77]; hence, we cannot use the Fredholm theory to infer the invertibility. For a résumé on these topics one can consult [21]. Verchota in [131] solved this problem showing the fundamental idea that the Rellich identities are the appropriate substitutes of compactness in the case of Lipschitz domains. Here, we recall the Rellich identity for the Laplace equation.

Proposition 2.2.6 ([21]) *Let Ω be a bounded Lipschitz domain in $\mathbb{R}^d$. Let u be a function such that either*

(i) u is Lipschitz in $\overline{\Omega}$ and $\Delta u = 0$ in Ω,
or
(ii) u is Lipschitz in $\mathbb{R}^d \setminus \Omega$ and $\Delta u = 0$ in $\mathbb{R}^d \setminus \overline{\Omega}$ with $|u| = O(|\boldsymbol{x}|^{2-d})$
Let $\boldsymbol{\alpha}$ be a C^1-vector field in $\mathbb{R}^d$ with compact support. Then

$$\int_{\partial\Omega} (\boldsymbol{\alpha}\cdot\boldsymbol{n})\left|\frac{\partial u}{\partial \boldsymbol{n}}\right|^2 = \int_{\partial\Omega} (\boldsymbol{\alpha}\cdot\boldsymbol{n})\left|\frac{\partial u}{\partial \boldsymbol{t}}\right|^2 - 2\int_{\partial\Omega} \left(\boldsymbol{\alpha}\cdot\frac{\partial u}{\partial \boldsymbol{t}}\right)\frac{\partial u}{\partial \boldsymbol{n}}$$
$$+\begin{cases} \displaystyle\int_{\Omega} 2(\nabla\boldsymbol{\alpha}\nabla u\cdot\nabla u) - \operatorname{div}\boldsymbol{\alpha}|\nabla u|^2 & \textit{if } u \textit{ satisfies (i)} \\ \displaystyle\int_{\mathbb{R}^d\setminus\overline{\Omega}} 2(\nabla\boldsymbol{\alpha}\nabla u\cdot\nabla u) - \operatorname{div}\boldsymbol{\alpha}|\nabla u|^2 & \textit{if } u \textit{ satisfies (ii).} \end{cases} \tag{2.19}$$

We have that there exists, as a consequence of the previous Rellich formula, a positive constant C depending only on the Lipschitz character of Ω such that

$$\frac{1}{C}\left\|\frac{\partial u}{\partial \boldsymbol{t}}\right\|_{L^2(\partial\Omega)} \leq \left\|\frac{\partial u}{\partial \boldsymbol{n}}\right\|_{L^2(\partial\Omega)} \leq C\left\|\frac{\partial u}{\partial \boldsymbol{t}}\right\|_{L^2(\partial\Omega)}. \tag{2.20}$$

For more details see [21].

In the proof of the invertibility of the operators $\lambda I - K^*_\Omega$ a crucial role is played by the following theorem.

Theorem 2.2.7 *For $0 \leq h \leq 1$ suppose that the family of operators A_h : $L^2(\mathbb{R}^{d-1}) \to L^2(\mathbb{R}^{d-1})$ satisfy*

(i) $\|A_h\varphi\|_{L^2(\mathbb{R}^{d-1})} \geq C\|\varphi\|_{L^2(\mathbb{R}^{d-1})}$, *where C is independent of h;*
(ii) $h \to A_h$ *is continuous in norm;*
(iii) $A_0 : L^2(\mathbb{R}^{d-1}) \to L^2(\mathbb{R}^{d-1})$ *is invertible.*

Then, $A_1 : L^2(\mathbb{R}^{d-1}) \to L^2(\mathbb{R}^{d-1})$ *is invertible.*

All these ingredients allow to prove the invertibility of the operators $\lambda I - K^*_\Omega$, where λ belongs to the range expressed by Proposition 2.2.5. These results are due to Verchota [131] (for $\lambda = \pm 1/2$) and Escauriaza et al. [75]. For invertibility results in regular domains see [78] (for C^1-domains) and [84] (for C^2-domains).

Theorem 2.2.8 ([75]) *Let* Ω *be a Lipschitz domain. The operator* $\lambda I - K^*_\Omega$ *is invertible on*

(i) $L^2_0(\partial\Omega)$ *if* $|\lambda| \geq \frac{1}{2}$;
(ii) $L^2(\partial\Omega)$ *if* $\lambda \in (-\infty, -\frac{1}{2}] \cup (\frac{1}{2}, \infty)$.

Proof This proof is contained in [21]. We just recall it for reader's convenience. We first prove the invertibility of the operators $\pm 1/2I + K^*_\Omega : L^2_0(\partial\Omega) \to L^2_0(\partial\Omega)$ because it is essentially based on inequalities (2.20) and Theorem 2.2.7. Then we prove the general case which is more involved.

Case $\pm 1/2I + K^*_\Omega$
Since $K_\Omega(1) = 1/2$ we have that, for all $f \in L^2(\partial\Omega)$,

$$\int_{\partial\Omega} K^*_\Omega f(\boldsymbol{x})\, d\sigma(\boldsymbol{x}) = \frac{1}{2}\int_{\partial\Omega} f(\boldsymbol{x})\, d\sigma(\boldsymbol{x}),$$

hence, $\pm 1/2I + K^*_\Omega$ maps $L^2_0(\partial\Omega)$ into $L^2_0(\partial\Omega)$. We define $u(\boldsymbol{x}) = S_\Omega f(\boldsymbol{x})$, where $f \in L^2_0(\partial\Omega)$, which satisfies conditions (i) and (ii) in Proposition 2.2.6. Moreover by the properties of single layer potentials on the boundary of Ω, we have that $\partial u/\partial \boldsymbol{t}$ is continuous across the boundary and the jump relation holds

$$\frac{\partial u}{\partial \boldsymbol{n}}\Big|_{\pm} = \Big(\pm\frac{1}{2}I + K^*_\Omega\Big) f.$$

Applying (2.20) in Ω and $\mathbb{R}^d \setminus \overline{\Omega}$ we obtain that

$$\begin{aligned} \frac{1}{C}\Big\|\Big(\frac{1}{2}I + K^*_\Omega\Big) f\Big\|_{L^2(\partial\Omega)} &\leq \Big\|\Big(\frac{1}{2}I - K^*_\Omega\Big) f\Big\|_{L^2(\partial\Omega)} \\ \Big\|\Big(\frac{1}{2}I - K^*_\Omega\Big) f\Big\|_{L^2(\partial\Omega)} &\leq C\Big\|\Big(\frac{1}{2}I + K^*_\Omega\Big) f\Big\|_{L^2(\partial\Omega)}. \end{aligned} \tag{2.21}$$

Since

$$f = \Big(\frac{1}{2}I + K^*_\Omega\Big) f + \Big(\frac{1}{2}I - K^*_\Omega\Big) f,$$

from (2.21) we have that

$$\| \left(\frac{1}{2}I + K_{\Omega}^{*}\right) f \|_{L^2(\partial\Omega)} \geq C \|f\|_{L^2(\partial\Omega)}. \qquad (2.22)$$

Localizing the situation, we can assume that $\partial\Omega$ is the graph of a Lipschitz function in order to simplify as much as possible the proof. Therefore $\partial\Omega = \{(\boldsymbol{x}', x_d) : x_d = \psi(\boldsymbol{x}')\}$, where $\psi : \mathbb{R}^{d-1} \to \mathbb{R}$ is a Lipschitz function, see Definition 1.1.1.

We consider the Lipschitz graph corresponding to $h\psi$, that is,

$$\partial\Omega_h = \left\{(\boldsymbol{x}', x_d) : x_d = h\psi(\boldsymbol{x}')\right\}, \qquad 0 \leq h \leq 1,$$

to prove that the operator $A = (1/2)I + K_{\Omega}^{*}$ is invertible. In particular, we indicate with $K_{\Omega_h}^{*}$ and A_h the operator associated to the boundary $\partial\Omega_h$. Therefore, from the definition, it follows that $A_0 = (1/2)I$ and $A_1 = A$. The operators A_h are continuous in norm as a function of h. Hence, from the inequality in (2.22) we have that $\|A_h f\|_{L^2(\partial\Omega_h)} \geq C\|f\|_{L^2(\partial\Omega_h)}$, since the constant C is independent of h but depends only on the Lipschitz character of Ω. Applying the continuity method of Theorem 2.2.7 we find that $1/2I + K_{\Omega}^{*}$ is invertible on $L_0^2(\partial\Omega)$. Next, we prove that $1/2I + K_{\Omega}^{*}$ is invertible on $L^2(\partial\Omega)$ showing that the operator is onto on $L^2(\partial\Omega)$. By duality argument, since $K_{\Omega}(1) = 1/2$, for all $f \in L^2(\partial\Omega)$ we get

$$\int_{\partial\Omega} \left(\frac{1}{2}I + K_{\Omega}^{*}\right) f \, d\sigma(\boldsymbol{x}) = \int_{\partial\Omega} f \, d\sigma(\boldsymbol{x}),$$

hence, $1/2I + K_{\Omega}^{*}$ maps $L^2(\partial\Omega)$ into $L^2(\partial\Omega)$. For $g \in L^2(\partial\Omega)$ we consider

$$g = g - c\left(\frac{1}{2}I + K_{\Omega}^{*}\right)(1) + c\left(\frac{1}{2}I + K_{\Omega}^{*}\right)(1),$$

where

$$c = \frac{1}{|\partial\Omega|} \int_{\partial\Omega} g \, d\sigma(\boldsymbol{x}).$$

Defining $g_0 := g - c(1/2I + K_{\Omega}^{*})(1)$, since

$$\int_{\partial\Omega} (\frac{1}{2}I + K_{\Omega}^{*})(1) \, d\sigma(\boldsymbol{x}) = |\partial\Omega|,$$

we have that $g_0 \in L_0^2(\partial\Omega)$. Let $f_0 \in L_0^2(\partial\Omega)$ be such that

$$\left(\frac{1}{2}I + K_\Omega^*\right) f_0 = g_0.$$

Then, defining $f := f_0 + c$ we find that

$$\left(\frac{1}{2}I + K_\Omega^*\right) f = g_0 + c\left(\frac{1}{2}I + K_\Omega^*\right)(1) = g.$$

This means that $1/2I + K_\Omega^*$ is onto in $L^2(\partial\Omega)$.

For the operator $-1/2I + K_\Omega^*$ we can follow the same argument both for the case $L_0^2(\partial\Omega)$ and $L^2(\partial\Omega)$.

Next, we assume that $|\lambda| > 1/2$. To prove the invertibility of the operators in the general case we use the Rellich identity. Let $f \in L^2(\partial\Omega)$, c_0 a fixed positive number and set $u(\boldsymbol{x}) = S_\Omega f(\boldsymbol{x})$. Let $\boldsymbol{\alpha}$ be a vector field with support in the set dist($\boldsymbol{x}$,$\partial\Omega$)$< 2c_0$, $\forall \boldsymbol{x} \in \partial\Omega$, such that $\boldsymbol{\alpha} \cdot \boldsymbol{n} \geq \delta$, for some $\delta > 0$. Therefore, from the Rellich identity (2.2.6), we have

$$\begin{aligned}\int_{\partial\Omega} (\boldsymbol{\alpha} \cdot \boldsymbol{n})\left|\frac{\partial u}{\partial \boldsymbol{n}}\right|^2 &= \int_{\partial\Omega} (\boldsymbol{\alpha} \cdot \boldsymbol{n})\left|\frac{\partial u}{\partial \boldsymbol{t}}\right|^2 - 2\int_{\partial\Omega} (\boldsymbol{\alpha} \cdot \frac{\partial u}{\partial \boldsymbol{t}})\frac{\partial u}{\partial \boldsymbol{n}} \\ &+ \int_\Omega 2(\nabla\boldsymbol{\alpha}\nabla u \cdot \nabla u) - \operatorname{div}\boldsymbol{\alpha}\, |\nabla u|^2.\end{aligned} \tag{2.23}$$

Observe that on $\partial\Omega$

$$\frac{\partial u}{\partial \boldsymbol{n}}\bigg|_- = \left(-\frac{1}{2}I + K_\Omega^*\right) f = \left(\lambda - \frac{1}{2}\right) f - (\lambda I - K_\Omega^*) f.$$

Since $\boldsymbol{\alpha} = (\boldsymbol{\alpha} \cdot \boldsymbol{n})\boldsymbol{n} + \sum_{k=1}^{d-1}(\boldsymbol{\alpha} \cdot t_k)t_k$ and $\nabla S f(\boldsymbol{x})|_+ = 1/2\boldsymbol{n} f + \mathcal{K} f$, where

$$\mathcal{K} f(\boldsymbol{x}) = \frac{1}{\omega_d}\, p.v. \int_{\partial\Omega} \frac{\boldsymbol{x} - \boldsymbol{y}}{|\boldsymbol{x} - \boldsymbol{y}|^d} f(\boldsymbol{y})\, d\sigma(\boldsymbol{y}),$$

we find that

$$\begin{aligned}(\nabla u \cdot \boldsymbol{\alpha}) &= \frac{\partial u}{\partial \boldsymbol{n}}(\boldsymbol{\alpha} \cdot \boldsymbol{n}) + (\boldsymbol{\alpha} \cdot \frac{\partial u}{\partial \boldsymbol{t}}) \\ &= -\frac{1}{2}(\boldsymbol{\alpha} \cdot \boldsymbol{n}) f + K_{\boldsymbol{\alpha}} f,\end{aligned} \tag{2.24}$$

where

$$K_{\boldsymbol{\alpha}} f(\boldsymbol{x}) = \frac{1}{\omega_d}\, p.v. \int_{\partial\Omega} \frac{((\boldsymbol{x} - \boldsymbol{y}) \cdot \boldsymbol{\alpha}(\boldsymbol{x}))}{|\boldsymbol{x} - \boldsymbol{y}|^d} f(\boldsymbol{y})\, d\sigma(\boldsymbol{y}).$$

We also have

$$\begin{aligned}\int_{\Omega} |\nabla u|^2 \, d\boldsymbol{x} &= \int_{\partial\Omega} u \frac{\partial u}{\partial \boldsymbol{n}}\Big|_{-} \, d\sigma(\boldsymbol{x}) \\ &= \int_{\partial\Omega} \mathcal{S}_\Omega f \left[\left(\lambda - \frac{1}{2}\right) f - (\lambda I - \mathcal{K}_\Omega^*) f \right] d\sigma(\boldsymbol{x}).\end{aligned}$$

By using the following integral identity obtained by multiplying (2.24) for $\partial u / \partial \boldsymbol{n}$, that is,

$$\begin{aligned}-2 \int_{\partial\Omega} \left(\boldsymbol{\alpha} \cdot \frac{\partial u}{\partial \boldsymbol{t}}\right) \frac{\partial u}{\partial \boldsymbol{n}} \, d\sigma(\boldsymbol{x}) =& -2 \int_{\partial\Omega} \frac{\partial u}{\partial \boldsymbol{n}} \left[-\frac{1}{2} (\boldsymbol{\alpha} \cdot \boldsymbol{n}) f + K_{\boldsymbol{\alpha}} f \right] d\sigma(\boldsymbol{x}) \\ &+ 2 \int_{\partial\Omega} (\boldsymbol{\alpha} \cdot \boldsymbol{n}) \left| \frac{\partial u}{\partial \boldsymbol{n}} \right|^2 d\sigma(\boldsymbol{x}),\end{aligned}$$

we get from the Rellich formula (2.23) that

$$\begin{aligned}\frac{1}{2} \int_{\partial\Omega} (\boldsymbol{\alpha} \cdot \boldsymbol{n}) \left| \frac{\partial u}{\partial \boldsymbol{n}} \right|^2 d\sigma(\boldsymbol{x}) =& -\frac{1}{2} \int_{\partial\Omega} (\boldsymbol{\alpha} \cdot \boldsymbol{n}) \left| \frac{\partial u}{\partial \boldsymbol{t}} \right|^2 d\sigma(\boldsymbol{x}) \\ &+ \int_{\partial\Omega} \frac{\partial u}{\partial \boldsymbol{n}} \left[-\frac{1}{2} (\boldsymbol{\alpha} \cdot \boldsymbol{n}) f + K_{\boldsymbol{\alpha}} f \right] d\sigma(\boldsymbol{x}) \\ &- \int_{\Omega} \left[(\nabla \boldsymbol{\alpha} \nabla u, \nabla u) + \frac{1}{2} \operatorname{div} \boldsymbol{\alpha} \, |\nabla u|^2 \right] d\sigma(\boldsymbol{x}).\end{aligned}$$

Thus it holds

$$\begin{aligned}&\frac{1}{2} \left(\lambda - \frac{1}{2}\right)^2 \int_{\partial\Omega} (\boldsymbol{\alpha} \cdot \boldsymbol{n}) f^2 \, d\sigma(\boldsymbol{x}) \\ &\leq \int_{\partial\Omega} \left[-\frac{1}{2} (\boldsymbol{\alpha} \cdot \boldsymbol{n}) f + K_{\boldsymbol{\alpha}} f \right] \left[\left(\lambda - \frac{1}{2}\right) f - (\lambda I - \mathcal{K}_\Omega^*) f \right] d\sigma(\boldsymbol{x}) \\ &\quad + C \|f\|_{L^2(\partial\Omega)} \left(\|\mathcal{S}_\Omega f\|_{L^2(\partial\Omega)} + \|(\lambda I - \mathcal{K}_\Omega^*) f\|_{L^2(\partial\Omega)} \right) \\ &\quad + C \|\mathcal{S}_\Omega f\|_{L^2(\partial\Omega)} \|(\lambda I - \mathcal{K}_\Omega^*) f\|_{L^2(\partial\Omega)} + C \|(\lambda I - \mathcal{K}_\Omega^*) f\|^2_{L^2(\partial\Omega)},\end{aligned}$$

where the constant C depends on the Lipschitz character of Ω and λ. By the multiplication of the terms in the right-hand side integral, we get

$$\frac{1}{2}\left(\lambda^2 - \frac{1}{4}\right) \int\limits_{\partial\Omega} (\boldsymbol{\alpha} \cdot \boldsymbol{n}) f^2 \, d\sigma(\boldsymbol{x}) \le \left(\lambda - \frac{1}{2}\right) \int\limits_{\partial\Omega} f K_{\boldsymbol{\alpha}} f \, d\sigma(\boldsymbol{x})$$
$$+ C\|f\|_{L^2(\partial\Omega)} \left(\|S_\Omega f\|_{L^2(\partial\Omega)} + \|(\lambda I - K_\Omega^*) f\|_{L^2(\partial\Omega)}\right)$$
$$+ C\|S_\Omega f\|_{L^2(\partial\Omega)} \|(\lambda I - K_\Omega^*) f\|_{L^2(\partial\Omega)} + C\|(\lambda I - K_\Omega^*) f\|^2_{L^2(\partial\Omega)}.$$

Denoting with $K_{\boldsymbol{\alpha}}^*$ the adjoint operator in $L^2(\partial\Omega)$ of $K_{\boldsymbol{\alpha}}$ we find

$$K_{\boldsymbol{\alpha}}^* + K_{\boldsymbol{\alpha}} = R_{\boldsymbol{\alpha}} f = \frac{1}{\omega_d} p.v. \int\limits_{\partial\Omega} \frac{[(\boldsymbol{x} - \boldsymbol{y}) \cdot (\boldsymbol{\alpha}(\boldsymbol{x}) - \boldsymbol{\alpha}(\boldsymbol{y}))]}{|\boldsymbol{x} - \boldsymbol{y}|^d} f(\boldsymbol{y}) \, d\sigma(\boldsymbol{y}).$$

By duality, we have

$$\int\limits_{\partial\Omega} f K_{\boldsymbol{\alpha}} f \, d\sigma(\boldsymbol{x}) = \frac{1}{2} \int\limits_{\partial\Omega} f R_{\boldsymbol{\alpha}} f \, d\sigma(\boldsymbol{x}).$$

Since $|\lambda| > 1/2$ and $\boldsymbol{\alpha} \cdot \boldsymbol{n} \ge \delta > 0$, the norm $\|f\|_{L^2(\partial\Omega)}$ in the left-hand side can be estimated by

$$\|f\|_{L^2(\partial\Omega)} \le C \left(\|(\lambda I - K_\Omega^*) f\|_{L^2(\partial\Omega)} + \|S_\Omega f\| + \|R_\alpha f\|_{L^2(\partial\Omega)}\right). \tag{2.25}$$

Since S_Ω and $R_{\boldsymbol{\alpha}}$ are compact in $L^2(\partial\Omega)$, we conclude from the above estimate that $\lambda I - K_\Omega^*$ has a closed range.

Next, we have to prove the surjectivity of the operator $\lambda I - K_{\partial\Omega}^*$ in $L^2(\partial\Omega)$. From this result and the injectivity proved in Theorem 2.2.5 the invertibility follows. Indeed, by contradiction, suppose that for some λ real, $|\lambda| > 1/2$, the operator $\lambda I - K_\Omega^*$ is not invertible in $L^2(\partial\Omega)$. Then, the intersection of the spectrum of K_Ω^* and the set $\{\lambda \in \mathbb{R} : |\lambda| > 1/2\}$ is not empty and so there exists a real number λ_0 that belongs to this intersection and it is a boundary point of the set. To get a contradiction we have to show that $\lambda_0 I - K_\Omega^*$ is invertible. We know that $\lambda_0 I - K_\Omega^*$ is injective and by (2.25) has a closed range. This means that there exists a constant C such that for all $f \in L^2(\partial\Omega)$,

$$\|f\|_{L^2(\partial\Omega)} \le C\|(\lambda_0 I - K_\Omega^*) f\|_{L^2(\partial\Omega)}. \tag{2.26}$$

Since λ_0 is a boundary point of the intersection of the spectrum of K_Ω^* and the real line there exists a real sequence λ_k with $|\lambda_k| > 1/2$, $\lambda_k \to \lambda_0$, as $k \to \infty$, such that $\lambda_k I - K_\Omega^*$ is invertible on $L^2(\partial\Omega)$. Therefore, given $g \in L^2(\partial\Omega)$ there exists a unique $f_k \in L^2(\partial\Omega)$ such that $(\lambda_k I - K_\Omega^*) f_k = g$. If $\{\|f_k\|_{L^2(\partial\Omega)}\}$ has a bounded subsequence then there exists another subsequence that converges weakly to some $f_0 \in L^2(\partial\Omega)$ and it holds

$$\int_{\partial\Omega} h(\lambda_0 I - K^*_\Omega) f_0 \, d\sigma(\mathbf{x}) = \lim_{k\to+\infty} \int_{\partial\Omega} f_k(\lambda_0 I - K_\Omega) h \, d\sigma(\mathbf{x})$$

$$= \lim_{k\to+\infty} \int_{\partial\Omega} h(\lambda_0 I - K^*_\Omega) f_k \, d\sigma(\mathbf{x}) = \int_{\partial\Omega} gh \, d\sigma(\mathbf{x}).$$

Therefore $(\lambda_0 I - K^*_\Omega) f_0 = g$. On the other hand, we may assume that $\| f_k \|_{L^2(\partial\Omega)} = 1$ and $(\lambda_0 I - K^*_\Omega) f_k$ converges to zero in $L^2(\partial\Omega)$. However, from (2.26), we have that

$$\begin{aligned} 1 = \| f_k \|_{L^2(\partial\Omega)} &\le C \| (\lambda_0 I - K^*_\Omega) f_k \|_{L^2(\partial\Omega)} \\ &\le C \, |\lambda - \lambda_k| + C \| (\lambda_k I - K^*_\Omega) f_k \|_{L^2(\partial\Omega)}. \end{aligned}$$

Since the right-hand side converges to zero as $k \to \infty$, we get a contradiction. This means that, for each λ real with $|\lambda| > 1/2$, $\lambda I - K^*_\Omega$ is invertible. □

Remark 2.2.9 The invertibility of the operator $1/2I + K_\Omega$ follows exploiting the Banach's closed range theorem starting from the result for $1/2I + K^*_\Omega$. In particular, the result follows from the fact that $1/2I + K^*_\Omega$ has closed and dense range in $L^2(\partial\Omega)$. For more details see [131].

2.3 The Scalar Problem

Thanks to the instruments introduced in the preliminary section, we are now ready to analyze the boundary value problem

$$\begin{cases} \Delta u = 0 & \text{in } \mathbb{R}^d_- \setminus C \\ \dfrac{\partial u}{\partial \boldsymbol{n}} = g & \text{on } \partial C \\ \dfrac{\partial u}{\partial x_d} = 0 & \text{on } \mathbb{R}^{d-1} \\ u \to 0 & \text{as } |\mathbf{x}| \to +\infty, \end{cases} \tag{2.27}$$

where C is the cavity, g is a function defined on ∂C and $d \ge 3$.

In particular, we establish the well-posedness of the problem and provide an integral representation formula for any bounded Lipschitz domain C contained in the half-space. Only in the last section of this chapter, making the smallness assumption on the cavity, we find the asymptotic expansion of the solution of (2.27).

In this chapter we also prove that the well-posedness of the boundary value problem (2.27) can be obtained by weighted Sobolev spaces, see, for example,

[28–30, 32]. Here, this different approach serves to become familiar with the weak formulation in the half-space through weighted functional spaces since the same procedure will be applied in Chap. 3 in the more difficult context of linear elasticity. The choice to work with the weak formulation of (2.27) lies in the fact that the application of Lax–Milgram theorem allows to provide quantitative estimates of the solution.

2.3.1 Well-Posedness

The setting of a half-space and, in general, of an unbounded domain with unbounded boundary is more complicated than the case of exterior or bounded domains to get the well-posedness of the boundary value problem. Indeed, it is necessary to know the behavior of both solution decay at infinity and its integrability on the boundary.

From the particular symmetry of the half-space and the Laplace operator, the well-posedness of (2.27) follows extending the problem to the whole space, specifically to an exterior domain in $\mathbb{R}^d$.

Given a bounded Lipschitz domain $C \subset \mathbb{R}^d_-$ and the function $g : \partial C \to \mathbb{R}$, we define

$$\widetilde{C} := \{(\boldsymbol{x}', x_d) : (\boldsymbol{x}', -x_d) \in C\},$$

see Fig. 2.1, and $G : \partial C \cup \partial\widetilde{C} \to \mathbb{R}$ as

$$G(\boldsymbol{x}) := \begin{cases} g(\boldsymbol{x}) & \text{if } \boldsymbol{x} \in \partial C \\ g(\widetilde{\boldsymbol{x}}) & \text{if } \boldsymbol{x} \in \partial\widetilde{C}. \end{cases} \tag{2.28}$$

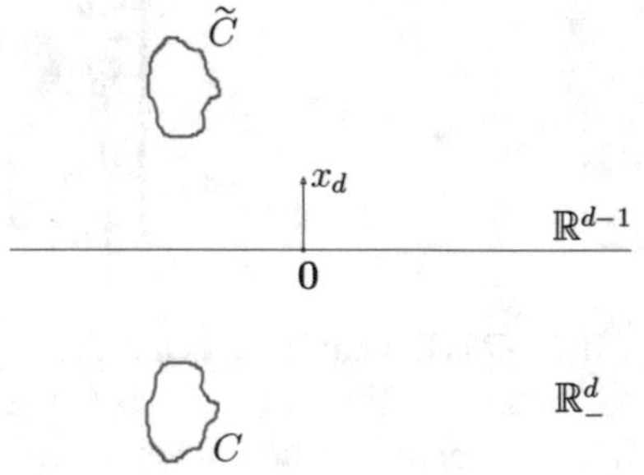

Fig. 2.1 Reflection of the geometry

Then, we consider the following exterior problem:

$$\begin{cases} \Delta U = 0 & \text{in } \mathbb{R}^d \setminus \left(C \cup \widetilde{C}\right) \\ \dfrac{\partial U}{\partial \boldsymbol{n}} = G & \text{on } \partial C \cup \partial \widetilde{C} \\ U \to 0 & \text{as } |\boldsymbol{x}| \to +\infty. \end{cases} \tag{2.29}$$

We first show the well-posedness of this problem.

Lemma 2.3.1 *There exists a unique solution for* (2.29).

Proof For $\Lambda := C \cup \widetilde{C}$, let $R > 0$ be such that $\Lambda \subset B_R(\mathbf{0})$ and set $\Omega_R := B_R(\mathbf{0}) \setminus \Lambda$.

Uniqueness Given two solutions, U_1 and U_2, to problem (2.29), the difference $W := U_1 - U_2$ solves the corresponding homogeneous problem, i.e.,

$$\begin{cases} \Delta W = 0 & \text{in } \mathbb{R}^d \setminus \overline{\Lambda} \\ \dfrac{\partial W}{\partial \boldsymbol{n}} = 0 & \text{on } \partial \Lambda \\ W \to 0 & \text{as } |\boldsymbol{x}| \to +\infty. \end{cases} \tag{2.30}$$

Multiplying equation $\Delta W = 0$ by W and integrating over the domain $\Omega_R = B_R(\mathbf{0}) \setminus \Lambda$, we get

$$\begin{aligned} 0 &= \int_{\Omega_R} W(\boldsymbol{x}) \Delta W(\boldsymbol{x}) d\boldsymbol{x} \\ &= \int_{\partial \Omega_R} W(\boldsymbol{x}) \frac{\partial}{\partial \boldsymbol{n}} W(\boldsymbol{x}) d\sigma(\boldsymbol{x}) - \int_{\Omega_R} \left|\nabla W(\boldsymbol{x})\right|^2 d\boldsymbol{x} \\ &= \int_{\partial B_R(\mathbf{0})} W(\boldsymbol{x}) \frac{\partial}{\partial \boldsymbol{n}} W(\boldsymbol{x}) d\sigma(\boldsymbol{x}) - \int_{\Omega_R} \left|\nabla W(\boldsymbol{x})\right|^2 d\boldsymbol{x}, \end{aligned}$$

using integration by parts and boundary conditions on $\partial \Lambda$. Next, utilizing the explicit behavior of the harmonic functions on exterior domains, as described in Theorem 2.2.2, we find

$$\left| \int_{\partial B_R(\mathbf{0})} W(\boldsymbol{x}) \frac{\partial}{\partial \boldsymbol{n}} W(\boldsymbol{x})\, d\sigma(\boldsymbol{x}) \right| \leq \frac{C}{R^{d-2}}.$$

Therefore, as $R \to \infty$, it follows that

$$\int_{\mathbb{R}^d \setminus \Lambda} \left|\nabla W(\boldsymbol{x})\right|^2 d\boldsymbol{x} = 0,$$

which implies $W = 0$.

Existence We represent U in (2.29) as a single layer potential with a potential $\psi \in L^2(\partial\Lambda)$ which has to be determined, i.e.,

$$S_\Lambda\psi(\boldsymbol{x}) = \int_{\partial\Lambda} \Gamma(\boldsymbol{x}-\boldsymbol{y})\psi(\boldsymbol{y})d\sigma(\boldsymbol{y}), \qquad \boldsymbol{x} \in \mathbb{R}^d \setminus \Lambda. \tag{2.31}$$

By the properties of single layer potentials, i.e., $S_\Lambda\psi$ is harmonic in $\mathbb{R}^d \setminus \partial\Lambda$ and $S_\Lambda\psi(\boldsymbol{x})=O(|\boldsymbol{x}|^{2-d})$ as $|\boldsymbol{x}| \to \infty$, see Sect. 2.2.2, we have that the layer potential in (2.31) satisfies the Laplace equation and the decay conditions at infinity in (2.29). Moreover, we recall that on $\partial\Lambda$ holds

$$\left.\frac{\partial S_\Lambda\psi}{\partial \boldsymbol{n}}(\boldsymbol{x})\right|_+ = \frac{1}{2}\psi + K_\Lambda^*\psi, \qquad \boldsymbol{x} \in \partial\Lambda.$$

Observe that $G \in L^2(\partial\Lambda)$, hence, from the injectivity result on $L^2(\partial\Lambda)$ of the operator $1/2I + K_\Lambda^*$, see Theorem 2.2.5, there exists a function $\psi \in L^2(\partial\Lambda)$ such that

$$\left(\frac{1}{2}I + K_\Lambda^*\right)\psi(\boldsymbol{x}) = G(\boldsymbol{x}), \qquad \boldsymbol{x} \in \partial\Lambda. \tag{2.32}$$

This proves the existence. Hence the statement of the theorem follows.

□

Next, we are able to prove the well-posedness of (2.27).

Theorem 2.3.2 *There exists a unique solution for problem* (2.27). *This solution coincides with the restriction to the half-space* $\mathbb{R}^d_-$ *of the solution to* (2.29).

Proof We need to prove the equivalence between (2.27) and (2.29).

At this aim, we define $u(\boldsymbol{x}', x_d) := U(\boldsymbol{x}', x_d)\big|_{x_d<0}$, where U is the unique solution of (2.29). From the boundary value problem (2.29) for U and the definition (2.28) of the function G, we have

$$\begin{cases} \Delta u = 0 & \text{in } \mathbb{R}^d_- \setminus C \\ \dfrac{\partial u}{\partial \boldsymbol{n}} = g & \text{on } \partial C \\ u \to 0 & \text{as } |\boldsymbol{x}| \to +\infty. \end{cases}$$

We only need to verify that the normal derivative is null on the boundary of the half-space. The key point is to show that U is even with respect to the x_d-plane. We define

$$\bar{u}(\boldsymbol{x}', x_d) := U(\boldsymbol{x}', -x_d), \tag{2.33}$$

for $\boldsymbol{x} \in \mathbb{R}^d \setminus \left(C \cup \widetilde{C}\right)$. Then $\bar{u}$ solves the following problem:

$$\begin{cases} \Delta \bar{u} = 0 & \text{in } \mathbb{R}^d \setminus \left(C \cup \widetilde{C}\right) \\ \dfrac{\partial \bar{u}}{\partial \boldsymbol{n}} = G & \text{on } \partial C \cup \partial \widetilde{C} \\ \bar{u} \to 0 & \text{as } |\boldsymbol{x}| \to +\infty, \end{cases} \tag{2.34}$$

since G is even with respect to x_d. Moreover, on $\partial C \cap \partial \widetilde{C}$, we have

$$\frac{\partial \bar{u}}{\partial \boldsymbol{n}}(\boldsymbol{x}', x_d) = \frac{\partial U}{\partial \boldsymbol{n}}(\boldsymbol{x}', -x_d).$$

Problem (2.34) admits a unique solution $\bar{u}(\boldsymbol{x})$ as a consequence of Lemma 2.3.1, hence,

$$U(\boldsymbol{x}', -x_d) = \bar{u}(\boldsymbol{x}', x_d) = U(\boldsymbol{x}', x_d).$$

From this last result, we obtain

$$\frac{\partial \bar{u}}{\partial x_d}(\boldsymbol{x}', x_d) = \frac{\partial U}{\partial x_d}(\boldsymbol{x}', x_d) = -\frac{\partial U}{\partial x_d}(\boldsymbol{x}', -x_d),$$

from which it follows that the derivative of U with respect to x_d on $\{x_d = 0\}$ is zero. □

Well-Posedness via Weighted Sobolev Spaces

Another way to prove the well-posedness of problem (2.27) is to set the analysis in weighted Sobolev spaces see Sect. 2.1.1. As pointed out in [28], these spaces have some advantages. Indeed an optimal weighted Poincaré inequality holds and a description of both the behavior at infinity of functions and their gradients is possible.

The study of the well-posedness of

$$\begin{cases} \Delta u = 0 & \text{in } \mathbb{R}^d_- \setminus C \\ \dfrac{\partial u}{\partial \boldsymbol{n}} = g & \text{on } \partial C \\ \dfrac{\partial u}{\partial x_d} = 0 & \text{on } \mathbb{R}^{d-1}, \end{cases} \tag{2.35}$$

where we assume $g \in L^2(\partial C)$ will be accomplished by Lax–Milgram theorem in $H^1_w(\mathbb{R}^d_- \setminus \overline{C})$. For more general problems (Dirichlet, Neumann, and mixed problems) and results for Poisson equation in an exterior domain of a half-space via weighted Sobolev spaces, see [28].

Theorem 2.3.3 *For any $g \in L^2(\partial C)$, there exists a unique solution $u \in H^1_w(\mathbb{R}^d_- \setminus \overline{C})$ to* (2.35). *Moreover, there exists a positive constant c such that*

$$\|u\|_{H^1_w(\mathbb{R}^d_- \setminus \overline{C})} \leq cg.$$

Proof We assume, for the moment, u regular and the test functions v in $\mathcal{D}(\overline{\mathbb{R}^d_-} \setminus C)$. Then multiplying Δu in (2.35) for v, integrating in $\mathbb{R}^d_- \setminus C$ and using the boundary conditions, we obtain

$$\int_{\mathbb{R}^d_- \setminus \overline{C}} \nabla u \cdot \nabla v \, d\boldsymbol{x} = -\int_{\partial C} g v \, d\sigma(\boldsymbol{x}), \quad \forall v \in \mathcal{D}(\overline{\mathbb{R}^d_-} \setminus C).$$

The extension of this formulation to $H^1_w(\mathbb{R}^d_- \setminus \overline{C})$ comes from the density result in Remark 2.1.2. Therefore the boundary value problem (2.35) becomes

Find $u \in H^1_w(\mathbb{R}^d_- \setminus \overline{C})$ such that

$$a(u, v) = l(v), \quad \forall v \in H^1_w(\mathbb{R}^d_- \setminus \overline{C}), \tag{2.36}$$

where $a : H^1_w(\mathbb{R}^d_- \setminus \overline{C}) \times H^1_w(\mathbb{R}^d_- \setminus \overline{C}) \to \mathbb{R}$ is the bilinear form given by

$$a(u, v) = \int_{\mathbb{R}^d_- \setminus \overline{C}} \nabla u \cdot \nabla v \, d\boldsymbol{x}, \tag{2.37}$$

and $l : H^1_w(\mathbb{R}^d_- \setminus \overline{C}) \to \mathbb{R}$ is the linear functional given by

$$l(v) = -\int_{\partial C} g v \, d\sigma(\boldsymbol{x}). \tag{2.38}$$

The assertion of the theorem then follows by applying the Lax–Milgram theorem, once the continuity and coercivity of the bilinear form a and the continuity of the functional l are established. These are straightforward calculations but we add them for completeness.

Continuity and Coercivity of (2.36) From the Cauchy–Schwarz inequality we have

$$|a(u,v)| = \left| \int\limits_{\mathbb{R}^d_-\setminus\overline{C}} \nabla u \cdot \nabla v \, d\boldsymbol{x} \right| \le c \, \|\nabla u\|_{L^2(\mathbb{R}^d_-\setminus\overline{C})} \|\nabla v\|_{L^2(\mathbb{R}^d_-\setminus\overline{C})}$$

$$\le c \, \|u\|_{H^1_w(\mathbb{R}^d_-\setminus\overline{C})} \|v\|_{H^1_w(\mathbb{R}^d_-\setminus\overline{C})},$$

hence the continuity. Coercivity follows from the Poincaré-type inequality (2.5):

$$a(u,u) = \int\limits_{\mathbb{R}^d_-\setminus\overline{C}} |\nabla u|^2 \, d\boldsymbol{x} = \|\nabla u\|^2_{L^2(\mathbb{R}^d_-\setminus\overline{C})} \ge c \|u\|^2_{H^1_w(\mathbb{R}^d_-\setminus\overline{C})}.$$

Continuity of (2.38) Let us take $B_r^-(\boldsymbol{0}) = \{\boldsymbol{x} \in \mathbb{R}^d : \|\boldsymbol{x}\|^2 < r^2, \ x_d < 0\}$, with r sufficiently large such that to contain the cavity C. Then applying the trace theorem for bounded domains, we find

$$\left| -\int\limits_{\partial C} g v \, d\sigma(\boldsymbol{x}) \right| \le c \|g\|_{L^2(\partial C)} \|v\|_{L^2(\partial C)}$$

$$\le c \left(\left\| \frac{v}{\rho} \right\|_{L^2((B_r^-(\boldsymbol{0}))\setminus\overline{C})} + \|\nabla v\|_{L^2((B_r^-(\boldsymbol{0}))\setminus\overline{C})} \right)$$

$$\le c \, \|v\|_{H^1_w(\mathbb{R}^d_-\setminus\overline{C})}.$$

The conclusion now follows from the Lax–Milgram theorem.

□

2.3.2 Representation Formula

After proving the well-posedness of the boundary value problem (2.27), in this paragraph we derive an integral representation formula for the solution u to problem (2.27). It is based on the layer potentials defined in (2.14) where $\partial\Omega = \partial C$. Moreover, we define the layer potentials on the image cavity $\widetilde{C}$ as

$$\begin{aligned} \widetilde{S}_C\varphi(\boldsymbol{x}) &:= \int_{\partial C} \Gamma(\widetilde{\boldsymbol{x}} - \boldsymbol{y})\varphi(\boldsymbol{y})d\sigma(\boldsymbol{y}), && \boldsymbol{x} \in \mathbb{R}^d, \\ \widetilde{D}_C\varphi(\boldsymbol{x}) &:= \int_{\partial C} \frac{\partial}{\partial \boldsymbol{n}_{\boldsymbol{y}}} \Gamma(\widetilde{\boldsymbol{x}} - \boldsymbol{y})\varphi(\boldsymbol{y})d\sigma(\boldsymbol{y}), && \boldsymbol{x} \in \mathbb{R}^d \setminus \partial\widetilde{C}. \end{aligned} \tag{2.39}$$

They come from the reflection of the function φ with respect to the x_d-coordinate. For simplicity we refer to them as *image layer potentials*.

The following integral representation formula for the solution of (2.27) holds.

Theorem 2.3.4 *The solution u to problem* (2.27) *is such that*

$$u(\boldsymbol{x}) = S_C g(\boldsymbol{x}) - D_C f(\boldsymbol{x}) + \widetilde{S}_C g(\boldsymbol{x}) - \widetilde{D}_C f(\boldsymbol{x}), \qquad \boldsymbol{x} \in \mathbb{R}^d_- \setminus C, \tag{2.40}$$

where S_C, D_C *are defined in* (2.14), $\widetilde{S}_C, \widetilde{D}_C$ *in* (2.39), *g is the Neumann boundary condition in* (2.27), *and f is the trace of u on* ∂C, *which satisfies*

$$\left(\frac{1}{2}I + K_C + \widetilde{D}_C\right) f = S_C g + \widetilde{S}_C g. \tag{2.41}$$

Before proving Theorem 2.3.4, we first recall the definition of the Neumann function for the Laplace operator, see, for example, [93], in order to rewrite (2.40) in a compact form. The Neumann function $N = N(\boldsymbol{x}, \boldsymbol{y})$ is the solution to

$$\begin{cases} \Delta_{\boldsymbol{y}} N(\boldsymbol{x}, \boldsymbol{y}) = \delta_{\boldsymbol{x}}(\boldsymbol{y}) & \text{in } \mathbb{R}^d_- \\ \dfrac{\partial N}{\partial y_d}(\boldsymbol{x}, \boldsymbol{y}) = 0 & \text{on } \mathbb{R}^{d-1}, \end{cases}$$

where $\delta_{\boldsymbol{x}}(\boldsymbol{y})$ is the delta function with the center in $\boldsymbol{x} \in \mathbb{R}^d_-$. The solution of this problem has an explicit expression which can be inferred applying the classical method of images (see [76])

$$N(\boldsymbol{x}, \boldsymbol{y}) = \frac{1}{\omega_d(2-d)}\left(\frac{1}{|\boldsymbol{x}-\boldsymbol{y}|^{d-2}} + \frac{1}{|\widetilde{\boldsymbol{x}}-\boldsymbol{y}|^{d-2}}\right).$$

For simplicity of notation, we define $\kappa_d := 1/\omega_d(2-d)$. Utilizing the Neumann function N, the representation formula (2.40) can be equivalently written as

$$\begin{aligned} u(\boldsymbol{x}) &= \mathcal{N}(f, g)(\boldsymbol{x}) \\ &:= \int_{\partial C}\left[N(\boldsymbol{x}, \boldsymbol{y})g(\boldsymbol{y}) - \frac{\partial}{\partial \boldsymbol{n}_{\boldsymbol{y}}}N(\boldsymbol{x}, \boldsymbol{y})f(\boldsymbol{y})\right]d\sigma(\boldsymbol{y}), \quad \boldsymbol{x} \in \mathbb{R}^d_- \setminus C, \end{aligned} \tag{2.42}$$

which we now prove.

Proof of Theorem 2.3.4 Given $R, \varepsilon > 0$ such that $C \subset B_R(\boldsymbol{0})$ and $B_\varepsilon(\boldsymbol{x}) \subset \mathbb{R}^d_- \setminus C$, let

$$\Omega_{R,\varepsilon} := \left(\mathbb{R}^d_- \cap B_R(\boldsymbol{0})\right) \setminus \left(C \cup B_\varepsilon(\boldsymbol{x})\right).$$

We also define $\partial B^h_R(\boldsymbol{0})$ as the intersection of the half ball $B^-_R(\boldsymbol{0}) = \{\boldsymbol{x} \in \mathbb{R}^d, |\boldsymbol{x}|^2 < R^2, x_d < 0\}$ with the boundary of the half-space, and with $\partial B^b_R(\boldsymbol{0})$ the spherical cap (see Fig. 2.2). Applying second Green's identity to $N(\boldsymbol{x}, \cdot)$ and u in $\Omega_{R,\varepsilon}$, we get

Fig. 2.2 Domain $\Omega_{R,\varepsilon}$ used to obtain the integral representation formula (2.40). Courtesy of Aspri et al. [35]

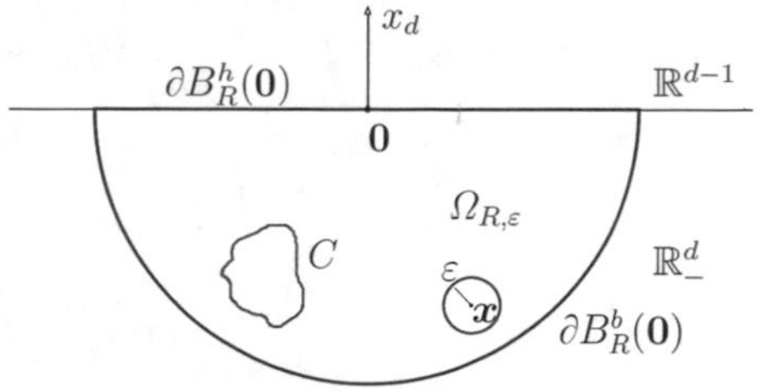

$$
\begin{aligned}
0 =& \int_{\Omega_{R,\varepsilon}} \left[N(\boldsymbol{x},\boldsymbol{y})\Delta u(\boldsymbol{y}) - u(\boldsymbol{y})\Delta_{\boldsymbol{y}} N(\boldsymbol{x},\boldsymbol{y})\right] d\boldsymbol{y} \\
=& \int_{\partial B_R^h(\mathbf{0})} \left[N(\boldsymbol{x},\boldsymbol{y})\frac{\partial u}{\partial y_d}(\boldsymbol{y}) - \frac{\partial}{\partial y_d} N(\boldsymbol{x},\boldsymbol{y}) u(\boldsymbol{y})\right] d\sigma(\boldsymbol{y}) \\
&+ \int_{\partial B_R^b(\mathbf{0})} \left[N(\boldsymbol{x},\boldsymbol{y})\frac{\partial u}{\partial \boldsymbol{n}_{\boldsymbol{y}}}(\boldsymbol{y}) - \frac{\partial}{\partial \boldsymbol{n}_{\boldsymbol{y}}} N(\boldsymbol{x},\boldsymbol{y}) u(\boldsymbol{y})\right] d\sigma(\boldsymbol{y}) \\
&+ \int_{\partial B_\varepsilon(\boldsymbol{x})} \left[\frac{\partial}{\partial \boldsymbol{n}_{\boldsymbol{y}}} N(\boldsymbol{x},\boldsymbol{y}) u(\boldsymbol{y}) - N(\boldsymbol{x},\boldsymbol{y})\frac{\partial u}{\partial \boldsymbol{n}_{\boldsymbol{y}}}(\boldsymbol{y})\right] d\sigma(\boldsymbol{y}) \\
&- \int_{\partial C} \left[N(\boldsymbol{x},\boldsymbol{y})\frac{\partial u}{\partial \boldsymbol{n}_{\boldsymbol{y}}}(\boldsymbol{y}) - \frac{\partial}{\partial \boldsymbol{n}_{\boldsymbol{y}}} N(\boldsymbol{x},\boldsymbol{y}) u(\boldsymbol{y})\right] d\sigma(\boldsymbol{y}) \\
=& I_1 + I_2 + I_3 - \mathcal{N}(f,g)(\boldsymbol{x}).
\end{aligned}
$$

We first note that the term I_1 is zero since both the normal derivative of the function N and u are zero above the boundary of the half-space.

Next, taking into account the behavior of harmonic functions in exterior domains, see (2.11), we find that

$$
\begin{aligned}
\left| \int_{\partial B_R^b(\mathbf{0})} \frac{\partial}{\partial \boldsymbol{n}_{\boldsymbol{y}}} N(\boldsymbol{x},\boldsymbol{y}) u(\boldsymbol{y})\, d\sigma(\boldsymbol{y}) \right| &\le \frac{C}{R^{2d-3}} \int_{\partial B_R^b(\mathbf{0})} d\sigma(\boldsymbol{y}) = \frac{C}{R^{d-2}}, \\
\left| \int_{\partial B_R^b(\mathbf{0})} N(\boldsymbol{x},\boldsymbol{y}) \frac{\partial u}{\partial \boldsymbol{n}_{\boldsymbol{y}}}(\boldsymbol{y})\, d\sigma(\boldsymbol{y}) \right| &\le \frac{C}{R^{2d-3}} \int_{\partial B_R^b(\mathbf{0})} d\sigma(\boldsymbol{y}) = \frac{C}{R^{d-2}},
\end{aligned}
$$

where C denotes a generic positive constant. Therefore I_2 tends to zero as $R \to +\infty$.

Finally, we decompose I_3 as

$$
I_3 = I_{31} - I_{32} = \int_{\partial B_\varepsilon(\boldsymbol{x})} \frac{\partial}{\partial \boldsymbol{n}_{\boldsymbol{y}}} N(\boldsymbol{x},\boldsymbol{y}) u(\boldsymbol{y})\, d\sigma(\boldsymbol{y}) - \int_{\partial B_\varepsilon(\boldsymbol{x})} N(\boldsymbol{x},\boldsymbol{y}) \frac{\partial u}{\partial \boldsymbol{n}_{\boldsymbol{y}}}(\boldsymbol{y}) d\sigma(\boldsymbol{y}).
$$

Using the expression of N, we derive

$$
\begin{aligned}
I_{31} &= \int_{\partial B_\varepsilon(\boldsymbol{x})} \frac{\partial}{\partial \boldsymbol{n}_{\boldsymbol{y}}} N(\boldsymbol{x}, \boldsymbol{y}) u(\boldsymbol{y}) d\sigma(\boldsymbol{y}) \\
&= u(\boldsymbol{x}) \int_{\partial B_\varepsilon(\boldsymbol{x})} \frac{\partial}{\partial \boldsymbol{n}_{\boldsymbol{y}}} N(\boldsymbol{x}, \boldsymbol{y}) d\sigma(\boldsymbol{y}) \\
&\quad + \int_{\partial B_\varepsilon(\boldsymbol{x})} [u(\boldsymbol{y}) - u(\boldsymbol{x})] \frac{\partial}{\partial \boldsymbol{n}_{\boldsymbol{y}}} N(\boldsymbol{x}, \boldsymbol{y}) d\sigma(\boldsymbol{y}),
\end{aligned}
$$

which tends to $u(\boldsymbol{x})$ as $\varepsilon \to 0$. In addition, it holds that

$$
\begin{aligned}
\left|I_{32}\right| &\le C \sup_{\boldsymbol{y} \in \partial B_\varepsilon(\boldsymbol{x})} \left|\frac{\partial u}{\partial \boldsymbol{n}_{\boldsymbol{y}}}\right| \int_{\partial B_\varepsilon(\boldsymbol{x})} \left|N(\boldsymbol{x}, \boldsymbol{y})\right| d\sigma(\boldsymbol{y}) \\
&\le C' \sup_{\boldsymbol{y} \in \partial B_\varepsilon(\boldsymbol{x})} \left|\frac{\partial u}{\partial \boldsymbol{n}_{\boldsymbol{y}}}\right| \left[\int_{\partial B_\varepsilon(\boldsymbol{x})} \frac{1}{\varepsilon^{d-2}} d\sigma(\boldsymbol{y}) + \int_{\partial B_\varepsilon(\boldsymbol{x})} \frac{1}{|\widetilde{\boldsymbol{x}} - \boldsymbol{y}|^{d-2}} d\sigma(\boldsymbol{y})\right].
\end{aligned}
$$

It is straightforward to see that $I_{32} \to 0$ as $\varepsilon \to 0$. Indeed, both the integrals tend to zero when ε goes to zero because the second one has a continuous kernel while the first one behaves as $O(\varepsilon)$. Putting together all the results, we obtain (2.42).

Finally, using the jump properties of the layer potentials, from equation (2.42) or in its more explicit form (2.40), we find that

$$
f(\boldsymbol{x}) = S_C g(\boldsymbol{x}) - \left(-\tfrac{1}{2} I + K_C\right) f(\boldsymbol{x}) - \widetilde{D}_C f(\boldsymbol{x}) + \widetilde{S}_C g(\boldsymbol{x}), \quad \boldsymbol{x} \in \partial C,
$$

where K_C is defined in (2.15). Thus, the trace f satisfies the integral equation

$$
\left(\frac{1}{2} I + K_C + \widetilde{D}_C\right) f = S_C g + \widetilde{S}_C g,
$$

which gives the assertion of the theorem. □

2.4 Spectral Analysis

In this section, for completeness, we want to study the invertibility of the operator $\frac{1}{2} I + K_C + \widetilde{D}_C$ in (2.41). We stress that in this way one would have another way to infer the well-posedness of (2.27). Indeed, rewriting (2.27) directly in terms of the integral equations (2.40) and (2.41), the existence of a unique solution would follow by the invertibility of the operator $\frac{1}{2} I + K_C + \widetilde{D}_C$. This argument will be used in the next chapter to prove the well-posedness of the elastic problem presented in the introduction.

To prove the invertibility of the operator $\frac{1}{2} I + K_C + \widetilde{D}_C$ we show that the following inclusion holds under suitable assumptions

$$
\sigma(K_C + \widetilde{D}_C) \subset (-1/2, 1/2] .
$$

In particular, noticing that the spectra of $K_C + \widetilde{D}_C$ and $K_C^* + \widetilde{D}_C^*$ are conjugate in $L^2(\partial C)$, we accomplish this task by finding the spectrum of $K_C^* + \widetilde{D}_C^*$ in $L^2(\partial C)$. For this purpose we follow the approach in the proof of Theorem 2.2.5.

The explicit expression of K_C^* is in (2.16). To find the L^2-adjoint of $\widetilde{D}_C$ is straightforward: indeed, given $\psi \in L^2(\partial C)$, we have

$$\begin{aligned}\int_{\partial C} \psi(\boldsymbol{x})\widetilde{D}_C\varphi(\boldsymbol{x})d\sigma(\boldsymbol{x}) &= \int_{\partial C} \psi(\boldsymbol{x})\left(\frac{1}{\omega_d}\int_{\partial C}\frac{(\boldsymbol{y}-\widetilde{\boldsymbol{x}})\cdot \boldsymbol{n}_{\boldsymbol{y}}}{|\widetilde{\boldsymbol{x}}-\boldsymbol{y}|^d}\varphi(\boldsymbol{y})d\sigma(\boldsymbol{y})\right)d\sigma(\boldsymbol{x})\\ &= \int_{\partial C} \varphi(\boldsymbol{y})\left(\frac{1}{\omega_d}\int_{\partial C}\frac{(\boldsymbol{y}-\widetilde{\boldsymbol{x}})\cdot \boldsymbol{n}_{\boldsymbol{y}}}{|\widetilde{\boldsymbol{x}}-\boldsymbol{y}|^d}\psi(\boldsymbol{x})\,d\sigma(\boldsymbol{x})\right)d\sigma(\boldsymbol{y}),\end{aligned}$$

and thus

$$\widetilde{D}_C^*\varphi(\boldsymbol{x}) = \frac{1}{\omega_d}\int_{\partial C}\frac{(\boldsymbol{x}-\widetilde{\boldsymbol{y}})\cdot \boldsymbol{n}_{\boldsymbol{x}}}{|\widetilde{\boldsymbol{y}}-\boldsymbol{x}|^d}\varphi(\boldsymbol{y})d\sigma(\boldsymbol{y}). \tag{2.43}$$

Note that the kernel of the integral operator $\widetilde{D}_C^*$ is smooth on ∂C.

In addition, we observe that $\widetilde{D}_C^*$, for $\boldsymbol{x} \in \partial C$, can be seen as the normal derivative of an appropriate single layer potential.

Lemma 2.4.1 *Given $\varphi \in L^2(\partial C)$ we have that*

$$\widetilde{D}_C^*\varphi(\boldsymbol{x}) = \frac{\partial}{\partial \boldsymbol{n}_{\boldsymbol{x}}}\left(S_{\widetilde{C}}\widetilde{\varphi}(\boldsymbol{x})\right), \quad \boldsymbol{x} \in \partial C,$$

where $\widetilde{\varphi} \in L^2(\partial \widetilde{C})$ is defined by $\widetilde{\varphi}(\boldsymbol{x}) := \varphi(\widetilde{\boldsymbol{x}})$.

Proof Using the expression (2.43) of $\widetilde{D}_C^*$ and the identity

$$\nabla_{\boldsymbol{x}}\left(\frac{1}{(2-d)|\boldsymbol{x}-\boldsymbol{y}|^{d-2}}\right) = \frac{\boldsymbol{x}-\boldsymbol{y}}{|\boldsymbol{x}-\boldsymbol{y}|^d},$$

we find that

$$\widetilde{D}_C^*\varphi(\boldsymbol{x}) = \nabla_{\boldsymbol{x}}\left(\int_{\partial C}\frac{\kappa_d\,\varphi(\boldsymbol{y})}{|\widetilde{\boldsymbol{y}}-\boldsymbol{x}|^{d-2}}\,d\sigma(\boldsymbol{y})\right)\cdot \boldsymbol{n}_{\boldsymbol{x}},$$

where $\kappa_d = 1/\omega_d(2-d)$. Given $\varphi \in L^2(\partial C)$ and $\widetilde{\varphi} \in L^2(\partial \widetilde{C})$ as previously defined, we have

$$\begin{aligned}\int_{\partial C}\frac{\varphi(\boldsymbol{y})}{|\widetilde{\boldsymbol{y}}-\boldsymbol{x}|^{d-2}}d\sigma(\boldsymbol{y}) &= \int_{\partial \widetilde{C}}\frac{\varphi(\widetilde{\boldsymbol{z}})}{|\widetilde{\widetilde{\boldsymbol{z}}}-\boldsymbol{x}|^{d-2}}\,d\sigma(\boldsymbol{z})\\ &= \int_{\partial \widetilde{C}}\frac{\varphi(\widetilde{\boldsymbol{z}})}{|\boldsymbol{z}-\boldsymbol{x}|^{d-2}}\,d\sigma(\boldsymbol{z}) = \int_{\partial \widetilde{C}}\frac{\widetilde{\varphi}(\boldsymbol{z})}{|\boldsymbol{z}-\boldsymbol{x}|^{d-2}}\,d\sigma(\boldsymbol{z}),\end{aligned}$$

which gives the conclusion. □

As recalled in Theorems 2.2.5 and 2.2.8 the eigenvalues of K_C^* on $L^2(\partial C)$ lie in $(-1/2, 1/2]$. With the same approach, we now show that the same property is true for $K_C^* + \widetilde{D}_C^*$.

Theorem 2.4.2 *Let C be an open bounded domain with Lipschitz boundary. Then*

$$\sigma(K_C^* + \widetilde{D}_C^*) \subset (-1/2, 1/2] .$$

Proof of Theorem (2.4.2) Given $\varphi \in L^2(\partial C)$, let ψ be defined by $\psi := S_C\varphi + S_{\widetilde{C}}\widetilde{\varphi}$. By the jump properties of single layer potentials, see (2.2.4), we find that

$$\left.\frac{\partial \psi}{\partial \boldsymbol{n}}\right|_{\pm} = \left(\pm\tfrac{1}{2}I + K_C^* + \widetilde{D}_C^*\right)\varphi, \qquad \text{on} \quad \partial C,$$

and, as a consequence,

$$\left.\frac{\partial \psi}{\partial \boldsymbol{n}}\right|_{+} + \left.\frac{\partial \psi}{\partial \boldsymbol{n}}\right|_{-} = 2\left(K_C^* + \widetilde{D}_C^*\right)\varphi, \qquad \left.\frac{\partial \psi}{\partial \boldsymbol{n}}\right|_{+} - \left.\frac{\partial \psi}{\partial \boldsymbol{n}}\right|_{-} = \varphi. \tag{2.44}$$

From these relations, we deduce that

$$\begin{aligned}\left(\lambda I - K_C^* - \widetilde{D}_C^*\right)\varphi &= \lambda\left(\left.\frac{\partial \psi}{\partial \boldsymbol{n}}\right|_{+} - \left.\frac{\partial \psi}{\partial \boldsymbol{n}}\right|_{-}\right) - \frac{1}{2}\left(\left.\frac{\partial \psi}{\partial \boldsymbol{n}}\right|_{+} + \left.\frac{\partial \psi}{\partial \boldsymbol{n}}\right|_{-}\right)\\ &= \left(\lambda - \frac{1}{2}\right)\left.\frac{\partial \psi}{\partial \boldsymbol{n}}\right|_{+} - \left(\lambda + \frac{1}{2}\right)\left.\frac{\partial \psi}{\partial \boldsymbol{n}}\right|_{-}.\end{aligned}$$

If λ is an eigenvalue of $K_C^* + \widetilde{D}_C^*$ with eigenfunction φ, then

$$\left(\lambda - \frac{1}{2}\right)\left.\frac{\partial \psi}{\partial \boldsymbol{n}}\right|_{+} - \left(\lambda + \frac{1}{2}\right)\left.\frac{\partial \psi}{\partial \boldsymbol{n}}\right|_{-} = 0, \qquad \text{on} \quad \partial C.$$

Multiplying such relation by the function ψ and integrating over ∂C, we get

$$\left(\lambda - \frac{1}{2}\right)\int_{\partial C}\psi(\boldsymbol{x})\left.\frac{\partial \psi}{\partial \boldsymbol{n}}(\boldsymbol{x})\right|_{+} d\sigma(\boldsymbol{x}) - \left(\lambda + \frac{1}{2}\right)\int_{\partial C}\psi(\boldsymbol{x})\left.\frac{\partial \psi}{\partial \boldsymbol{n}}(\boldsymbol{x})\right|_{-} d\sigma(\boldsymbol{x}) = 0. \tag{2.45}$$

From the integration by parts, it follows that

$$\begin{aligned}\int_{\partial C}\psi(\boldsymbol{x})\left.\frac{\partial \psi}{\partial \boldsymbol{n}}(\boldsymbol{x})\right|_{-} d\sigma(\boldsymbol{x}) &= \int_C \psi(\boldsymbol{x})\Delta\psi(\boldsymbol{x})\,d\boldsymbol{x} + \int_C \left|\nabla\psi(\boldsymbol{x})\right|^2 d\boldsymbol{x}\\ &= \int_C \left|\nabla\psi(\boldsymbol{x})\right|^2 d\boldsymbol{x}.\end{aligned} \tag{2.46}$$

For the first integral in (2.45) we can follow the argument in the proof of Theorem 2.3.4. Precisely, given large $R > 0$, applying the Green's formula in $\Omega_R := \left(\mathbb{R}^d_- \cap B^-_R(\mathbf{0})\right) \setminus C$, we get

$$\begin{aligned}\int_{\partial C} \psi(\boldsymbol{x})\frac{\partial\psi}{\partial\boldsymbol{n}}(\boldsymbol{x})\Big|_+ d\sigma(\boldsymbol{x}) &= \int_{\partial B^h_R(\mathbf{0})} \psi(\boldsymbol{x})\frac{\partial\psi}{\partial x_d}(\boldsymbol{x})\,d\sigma(\boldsymbol{x}) + \int_{\partial B^b_R(\mathbf{0})} \psi(\boldsymbol{x})\frac{\partial\psi}{\partial\boldsymbol{n}}(\boldsymbol{x})\Big|_+ d\sigma(\boldsymbol{x})\\ &\quad - \int_{\Omega_R} \psi(\boldsymbol{x})\Delta\psi(\boldsymbol{x})\,d\boldsymbol{x} - \int_{\Omega_R} |\nabla\psi(\boldsymbol{x})|^2\,d\boldsymbol{x},\end{aligned}$$

where $\partial B^h_R(\mathbf{0})$ is the intersection of the half ball with the half-space and $\partial B^b_R(\mathbf{0})$ is the spherical cap. The quantity $\partial\psi/\partial x_d$ is identically zero on the boundary of the half-space since the kernel of the operator is the normal derivative of the Neumann function which, by hypothesis, is null on $\mathbb{R}^{d-1}$. Moreover, ψ is harmonic in Ω_R, so we infer

$$\int_{\partial C} \psi(\boldsymbol{x})\frac{\partial\psi}{\partial\boldsymbol{n}}(\boldsymbol{x})\Big|_+ d\sigma(\boldsymbol{x}) = \int_{\partial B^b_R(\mathbf{0})} \psi(\boldsymbol{x})\frac{\partial\psi}{\partial\boldsymbol{n}}(\boldsymbol{x})\Big|_+ d\sigma(\boldsymbol{x}) - \int_{\Omega_R} |\nabla\psi(\boldsymbol{x})|^2\,d\boldsymbol{x}.$$

From the asymptotic behavior of simple layer potentials (Sect. 2.2.2), i.e.,

$$|S_C\varphi| + |S_{\widetilde{C}}\varphi| = O(|\boldsymbol{x}|^{2-d}), \qquad |\nabla S_C\varphi| + |\nabla S_{\widetilde{C}}\varphi| = O(|\boldsymbol{x}|^{1-d}) \quad \text{as } |\boldsymbol{x}| \to \infty,$$

we obtain, for some $C > 0$,

$$\begin{aligned}\left|\int_{\partial B^b_R(\mathbf{0})} \psi(\boldsymbol{x})\frac{\partial\psi}{\partial\boldsymbol{n}}(\boldsymbol{x})\Big|_+ d\sigma(\boldsymbol{x})\right| &\le \int_{\partial B^b_R(\mathbf{0})} |\psi(\boldsymbol{x})|\left|\frac{\partial\psi}{\partial\boldsymbol{n}}(\boldsymbol{x})\Big|_+\right| d\sigma(\boldsymbol{x})\\ &\le \frac{C}{R^{2d-3}}\int_{\partial B^b_R(\mathbf{0})} d\sigma(\boldsymbol{x}) = \frac{1}{R^{d-2}}.\end{aligned}$$

Therefore, as $R \to +\infty$, we find

$$\int_{\partial C} \psi(\boldsymbol{x})\frac{\partial\psi}{\partial\boldsymbol{n}}\Big|_+ d\sigma(\boldsymbol{x}) = -\int_{\mathbb{R}^d_-\setminus\overline{C}} |\nabla\psi(\boldsymbol{x})|^2\,d\boldsymbol{x}. \tag{2.47}$$

Plugging (2.46) and (2.47) into (2.45), we get the identity

$$\left(\lambda - \frac{1}{2}\right)\int_{\mathbb{R}^d_-\setminus C} |\nabla\psi(\boldsymbol{x})|^2\,d\boldsymbol{x} + \left(\lambda + \frac{1}{2}\right)\int_C |\nabla\psi(\boldsymbol{x})|^2\,d\boldsymbol{x} = 0. \tag{2.48}$$

Defining the two numbers

$$A := \int_{\mathbb{R}^d_- \setminus C} |\nabla \psi(\boldsymbol{x})|^2 \, d\boldsymbol{x} \qquad \text{and} \qquad B := \int_C |\nabla \psi(\boldsymbol{x})|^2 \, d\boldsymbol{x},$$

we have that (2.48) can be rewritten as

$$(A + B)\lambda = \frac{1}{2}(A - B).$$

The coefficient of λ is non-zero. On the contrary, if $A + B = 0$ then $\nabla\psi = 0$ in $\mathbb{R}^d_-$ which means that $\psi \equiv 0$, hence, from the second equation in (2.44), we get $\varphi = 0$ in ∂C.

Therefore, solving with respect to λ, we finally get

$$\lambda = \frac{1}{2} \cdot \frac{A - B}{A + B} \in \left[-\frac{1}{2}, \frac{1}{2} \right]. \tag{2.49}$$

The value $\lambda = -1/2$ is not an eigenvalue for the operator $K_C^* + \widetilde{D}_C^*$. Indeed, in such a case, we would have

$$A = \int_{\mathbb{R}^d_- \setminus C} |\nabla \psi(\boldsymbol{x})|^2 \, d\boldsymbol{x} = 0,$$

and thus $\psi = 0$ in $\mathbb{R}^d_- \setminus C$. By definition of ψ, we deduce that $\psi = 0$ on ∂C and since ψ is harmonic in C, we get that $\psi = 0$ also in C. As before, by (2.44), this would imply that $\varphi = 0$ in ∂C. □

For completeness, let us observe that the value $\lambda = 1/2$ is an eigenvalue. Indeed, identity (2.49) implies that, for such value of λ,

$$B = \int_C |\nabla \psi(\boldsymbol{x})|^2 \, d\boldsymbol{x} = 0,$$

hence, ψ is constant in C. Normalizing $\psi = 1$ in C and following the same approach of Sect. 2.3.1, the function ψ in $\mathbb{R}^d_- \setminus C$ is given by the restriction of the solution U to the Dirichlet problem in the exterior domain $\mathbb{R}^d \setminus (C \cup \widetilde{C})$ with boundary data equal to 1. Then, by the second equation in (2.44), the function φ is the normal derivative of U at ∂C.

2.5 Asymptotic Expansion

In this section we obtain an asymptotic expansion of the solution of (2.27) on the boundary of the half-space when the diameter of the cavity C goes to zero. We employ the approach of Ammari and Kang, see, for example, [21], based on integral equations, i.e., we utilize the representation formula which we proved in the previous section, precisely in Theorem 2.3.4. We assume that the cavity C is given by

$$C_\varepsilon := C = z + \varepsilon\Omega,$$

where z is the center of the cavity ($z \in \mathbb{R}^3_-$), ε is the parameters that goes to zero, and Ω is a bounded Lipschitz domain containing the origin. Moreover, we assume that the cavity is not touching the boundary, i.e.,

$$\mathrm{dist}(z, \mathbb{R}^{d-1}) \geq \delta_0 > 0, \tag{2.50}$$

otherwise some constants which will appear in the asymptotic expansion could explode to infinity. See, for example, discussions in Chapter 4 of [21]. In any case, thinking to problem (2.27) as a simplified mathematical model of the Mogi elastic model presented in the introduction, we stress that (2.50) is a reasonable request. The solution to the direct problem depending on the parameter ε is denoted by u_ε. At the same time, we also use the same notation for the layer potentials appearing in Theorem 2.3.4, i.e.,

$$S_\varepsilon = S_{C_\varepsilon}, \quad D_\varepsilon = D_{C_\varepsilon}, \quad \widetilde{S}_\varepsilon = \widetilde{S}_{C_\varepsilon}, \quad \widetilde{D}_\varepsilon = \widetilde{D}_{C_\varepsilon}, \quad K_\varepsilon = K_{C_\varepsilon},$$

and for the trace of the solution u_ε on ∂C_ε, i.e., f_ε. Therefore, Eq. (2.40) can be rewritten as

$$u_\varepsilon = S_\varepsilon g - D_\varepsilon f_\varepsilon - \widetilde{D}_\varepsilon f_\varepsilon + \widetilde{S}_\varepsilon g. \tag{2.51}$$

Since $\boldsymbol{x} = \widetilde{\boldsymbol{x}}$, it follows that for $\boldsymbol{x} \in \mathbb{R}^{d-1}$

$$S_\varepsilon g(\boldsymbol{x}) = \int_{\partial C_\varepsilon} \Gamma(\boldsymbol{x} - \boldsymbol{y}) g(\boldsymbol{y})\, d\sigma(\boldsymbol{y}) = \int_{\partial C_\varepsilon} \Gamma(\widetilde{\boldsymbol{x}} - \boldsymbol{y}) g(\boldsymbol{y})\, d\sigma(\boldsymbol{y}) = \widetilde{S}_\varepsilon g(\boldsymbol{x}),$$

and

$$\begin{aligned} D_\varepsilon f_\varepsilon(\boldsymbol{x}) &= \int_{\partial C_\varepsilon} \frac{\partial}{\partial \boldsymbol{n}_y} \Gamma(\boldsymbol{x} - \boldsymbol{y}) f_\varepsilon(\boldsymbol{y})\, d\sigma(\boldsymbol{y}) = \int_{\partial C_\varepsilon} \frac{\partial}{\partial \boldsymbol{n}_y} \Gamma(\widetilde{\boldsymbol{x}} - \boldsymbol{y}) f_\varepsilon(\boldsymbol{y})\, d\sigma(\boldsymbol{y}) \\ &= \widetilde{D}_\varepsilon f_\varepsilon(\boldsymbol{x}). \end{aligned}$$

Hence, Eq. (2.51) becomes

$$\tfrac{1}{2} u_\varepsilon(\boldsymbol{x}) = S_\varepsilon g(\boldsymbol{x}) - D_\varepsilon f_\varepsilon(\boldsymbol{x}), \qquad \boldsymbol{x} \in \mathbb{R}^{d-1}. \tag{2.52}$$

Recalling that f_ε on ∂C_ε satisfies (2.41), i.e.,

$$\left(\tfrac{1}{2}I + K_\varepsilon + \widetilde{D}_\varepsilon\right) f_\varepsilon(\boldsymbol{x}) = S_\varepsilon g(\boldsymbol{x}) + \widetilde{S}_\varepsilon g(\boldsymbol{x}), \qquad \boldsymbol{x} \in \partial C_\varepsilon, \tag{2.53}$$

and putting together (2.52) and (2.53) we get the identity

$$\tfrac{1}{2} u_\varepsilon(\boldsymbol{x}) = J_1(\boldsymbol{x}) + J_2(\boldsymbol{x}), \qquad \boldsymbol{x} \in \mathbb{R}^{d-1}, \tag{2.54}$$

where

$$J_1(\boldsymbol{x}) := \int_{\partial C_\varepsilon} \Gamma(\boldsymbol{x} - \boldsymbol{y}) g(\boldsymbol{y})\, d\sigma(\boldsymbol{y}),$$

$$J_2(\boldsymbol{x}) := -\int_{\partial C_\varepsilon} \frac{\partial}{\partial \boldsymbol{n}_{\boldsymbol{y}}} \Gamma(\boldsymbol{x} - \boldsymbol{y}) \left(\tfrac{1}{2}I + K_\varepsilon + \widetilde{D}_\varepsilon\right)^{-1} \left(S_\varepsilon g + \widetilde{S}_\varepsilon g\right)(\boldsymbol{y})\, d\sigma(\boldsymbol{y}).$$

To get the asymptotic expansion at first order of (2.54) on $\mathbb{R}^{d-1}$, we analyze in details the dependence with respect to ε of such relation.

In the sequel, for any fixed value of $\varepsilon > 0$, given $h : \partial C_\varepsilon \to \mathbb{R}$, we use the notation $h^\dagger$ to represent a function $h^\dagger : \partial\Omega \to \mathbb{R}$ which is defined by

$$h^\dagger(\boldsymbol{\zeta}; \varepsilon) := h(\boldsymbol{z} + \varepsilon\, \boldsymbol{\zeta}), \qquad \boldsymbol{\zeta} \in \partial\Omega.$$

Let us now state one of the main results contained in [35].

Theorem 2.5.1 *Let us assume* (2.50). *There exists* ε_0 *such that for all* $\varepsilon \in (0, \varepsilon_0)$ *and* $g \in L^2(\partial C_\varepsilon)$ *such that* $g^\dagger$ *is independent on* ε, *at any* $\boldsymbol{x} \in \mathbb{R}^{d-1}$ *the following expansion holds*

$$\begin{aligned} u_\varepsilon(\boldsymbol{x}) &= 2\varepsilon^{d-1} \Gamma(\boldsymbol{x} - \boldsymbol{z}) \int_{\partial\Omega} g^\dagger(\boldsymbol{\zeta})\, d\sigma(\boldsymbol{\zeta}) \\ &+ 2\varepsilon^d \nabla\Gamma(\boldsymbol{x} - \boldsymbol{z}) \cdot \int_{\partial\Omega} \left\{ \boldsymbol{n}_\zeta \left(\tfrac{1}{2}I + K_\Omega\right)^{-1} S_\Omega g^\dagger(\boldsymbol{\zeta}) - \boldsymbol{\zeta} g^\dagger(\boldsymbol{\zeta}) \right\} d\sigma(\boldsymbol{\zeta}) + O(\varepsilon^{d+1}), \end{aligned} \tag{2.55}$$

where $O(\varepsilon^{d+1})$ *denotes a quantity uniformly bounded by* $C\varepsilon^{d+1}$ *with* $C = C(\delta_0)$ *which tends to infinity when* δ_0 *goes to zero.*

To prove this theorem we first show the following expansion for the operator $\left(\tfrac{1}{2}I + K_\varepsilon + \widetilde{D}_\varepsilon\right)^{-1}$.

Lemma 2.5.2 *We have*

$$\left(\tfrac{1}{2}I + K_\varepsilon + \widetilde{D}_\varepsilon\right)^{-1} \left(S_\varepsilon g + \widetilde{S}_\varepsilon g\right)(\boldsymbol{z} + \varepsilon\boldsymbol{\zeta}) = \varepsilon \left(\tfrac{1}{2}I + K_\Omega\right)^{-1} S_\Omega g^\dagger(\boldsymbol{\zeta}) + O(\varepsilon^{d-1}) \tag{2.56}$$

for ε *sufficiently small.*

Proof To get the statement of this Lemma, we need to analyze, separately, the terms $\left(\frac{1}{2}I + K_\varepsilon + \widetilde{D}_\varepsilon\right)$ and $S_\varepsilon + \widetilde{S}_\varepsilon$.

At the point $z + \varepsilon\boldsymbol{\zeta}$, where $\boldsymbol{\zeta} \in \partial\Omega$, we obtain

$$K_\varepsilon\varphi(z + \varepsilon\boldsymbol{\zeta}) = \frac{1}{\omega_d}\text{p.v.}\int_{\partial C_\varepsilon} \frac{(\boldsymbol{y} - z - \varepsilon\boldsymbol{\zeta}) \cdot \boldsymbol{n}_{\boldsymbol{y}}}{|z + \varepsilon\boldsymbol{\zeta} - \boldsymbol{y}|^d}\varphi(\boldsymbol{y})\, d\sigma(\boldsymbol{y})$$
$$= \frac{1}{\omega_d}\text{p.v.}\int_{\partial\Omega} \frac{(\boldsymbol{\eta} - \boldsymbol{\zeta}) \cdot \boldsymbol{n}_{\boldsymbol{\eta}}}{|\boldsymbol{\zeta} - \boldsymbol{\eta}|^d}\varphi^\dagger(\boldsymbol{\eta})\, d\sigma(\boldsymbol{\eta}) = K_\Omega\varphi^\dagger(\boldsymbol{\zeta}),$$

and

$$\widetilde{D}_\varepsilon\varphi(z + \varepsilon\boldsymbol{\zeta}) = \int_{\partial C_\varepsilon} \frac{\partial}{\partial \boldsymbol{n}_{\boldsymbol{y}}}\Gamma(\widetilde{z} + \varepsilon\widetilde{\boldsymbol{\zeta}} - \boldsymbol{y})\varphi(\boldsymbol{y})\, d\sigma(\boldsymbol{y})$$
$$= \varepsilon^{d-1}\int_{\partial\Omega} \frac{\partial}{\partial \boldsymbol{n}_{\boldsymbol{\eta}}}\Gamma(\widetilde{z} + \varepsilon\widetilde{\boldsymbol{\zeta}} - z - \varepsilon\boldsymbol{\eta})\varphi^\dagger(\boldsymbol{\eta})\, d\sigma(\boldsymbol{\eta})$$
$$= \varepsilon^{d-1}R_\varepsilon\varphi^\dagger(\boldsymbol{\zeta}),$$

where

$$R_\varepsilon\varphi^\dagger(\boldsymbol{\zeta}) := \int_{\partial\Omega} \frac{\partial}{\partial \boldsymbol{n}_{\boldsymbol{\eta}}}\Gamma\left(\widetilde{z} - z + \varepsilon(\widetilde{\boldsymbol{\zeta}} - \boldsymbol{\eta})\right)\varphi^\dagger(\boldsymbol{\eta})\, d\sigma(\boldsymbol{\eta}) \tag{2.57}$$

is uniformly bounded in ε.

Let us evaluate the term $S_\varepsilon + \widetilde{S}_\varepsilon$. We have

$$S_\varepsilon g(z + \varepsilon\boldsymbol{\zeta}) = \int_{\partial C_\varepsilon} \Gamma(z + \varepsilon\boldsymbol{\zeta} - \boldsymbol{y})g(\boldsymbol{y})d\sigma(\boldsymbol{y})$$
$$= \varepsilon\int_{\partial\Omega} \Gamma(\boldsymbol{\zeta} - \boldsymbol{\theta})g^\dagger(\boldsymbol{\theta})d\sigma(\boldsymbol{\theta}) = \varepsilon S_\Omega g^\dagger(\boldsymbol{\zeta}),$$

and

$$\widetilde{S}_\varepsilon g(z + \varepsilon\boldsymbol{\zeta}) = \int_{\partial C_\varepsilon} \Gamma\left(\widetilde{z} + \varepsilon\widetilde{\boldsymbol{\zeta}} - \boldsymbol{y}\right) g(\boldsymbol{y})d\sigma(\boldsymbol{y})$$
$$= \varepsilon^{d-1}\int_{\partial\Omega} \Gamma\left(\widetilde{z} - z + \varepsilon(\widetilde{\boldsymbol{\zeta}} - \boldsymbol{\theta})\right) g^\dagger(\boldsymbol{\theta})d\sigma(\boldsymbol{\theta})$$
$$= \varepsilon^{d-1}\Gamma(\widetilde{z} - z)\int_{\partial\Omega} g^\dagger(\boldsymbol{\theta})d\sigma(\boldsymbol{\theta}) + O(\varepsilon^d),$$

where we have used the zero-order expansion for Γ. Therefore, from these two last results we get that

$$\left(S_\varepsilon g + \widetilde{S}_\varepsilon g\right)(z + \varepsilon \boldsymbol{\zeta}) = \varepsilon S_\Omega g^\dagger(\boldsymbol{\zeta}) + O(\varepsilon^{d-1}).$$

To conclude, from (2.53) we have

$$\left(\frac{1}{2}I + K_\Omega\right)\left(I + \varepsilon^{d-1}\left(\frac{1}{2}I + K_\Omega\right)^{-1} R_\varepsilon\right) f^\dagger = \varepsilon S_\Omega g^\dagger(\boldsymbol{\zeta}) + O(\varepsilon^{d-1}).$$

From the continuous property of R_ε and the invertibility result of the operator $1/2I + K_\Omega$ as explained in Remark 2.2.9, we have

$$\left\| \left(\frac{1}{2}I + K_\Omega\right)^{-1} R_\varepsilon \right\| \leq C,$$

where $C > 0$ is independent from ε. On the other hand, choosing $\varepsilon_0^{d-1} = 1/2C$, it follows that for all $\varepsilon \in (0, \varepsilon_0)$ we have that $I + \varepsilon^{d-1}\left(\frac{1}{2}I + K_\Omega\right)^{-1} R_\varepsilon$ is invertible and

$$\left(I + \varepsilon^{d-1}\left(\frac{1}{2}I + K_\Omega\right)^{-1} R_\varepsilon\right)^{-1} = I + O(\varepsilon^{d-1}).$$

Therefore

$$f^\dagger = \varepsilon\left(\frac{1}{2}I + K_\Omega\right)^{-1} S_\Omega g^\dagger(\boldsymbol{\zeta}) + O(\varepsilon^{d-1}).$$

□

Remark 2.5.3 If the domain C_ε is more regular, at least a C^1-domain, we have compactness of the operators K_ε and K_ε^*. Therefore we can prove the asymptotic expansion of the operator $\left(\frac{1}{2}I + K_\varepsilon + \widetilde{D}_\varepsilon\right)^{-1}$ in an alternative way. In fact, since $K_\varepsilon + \widetilde{D}_\varepsilon$ is compact and its spectrum is contained in $(-1/2, 1/2]$, there exists $\delta > 0$ such that

$$\sigma\left(K_\varepsilon + \widetilde{D}_\varepsilon\right) \subset (-1/2 + \delta, 1/2].$$

Therefore it follows that the operator

$$A_\varepsilon := \frac{1}{2}I - K_\varepsilon - \widetilde{D}_\varepsilon$$

has $\sigma(A_\varepsilon) \subset [0, 1 - \delta)$, i.e., it has spectral radius strictly smaller than 1. As a consequence, taking the powers of the operator A_ε one finds

$$\|A_\varepsilon^h\| \leq 1 \quad \forall h \quad \text{and} \quad \|A_\varepsilon^{h_0}\| < 1 \quad \text{for some } h_0. \tag{2.58}$$

Using the Neumann series, the inverse operator of $I - A_\varepsilon = \frac{1}{2}I + K_\varepsilon + \widetilde{D}_\varepsilon$ can be represented by

$$(I - A_\varepsilon)^{-1} = \sum_{h=0}^{+\infty} A_\varepsilon^h = \sum_{h=0}^{+\infty} \left(\frac{1}{2}I - K_\varepsilon - \widetilde{D}_\varepsilon\right)^h.$$

Taking R_ε of the proof of Lemma (2.5.2), see (2.57), we highlight the terms in A_ε^h which do not contain ε and that one of order $d-1$, that is,

$$A_\varepsilon^h = \left(\frac{1}{2}I - K_\Omega\right)^h - \varepsilon^{d-1} E_{h,\varepsilon},$$

where

$$E_{h,\varepsilon} = \sum_{j=1}^{h} A_\varepsilon \cdots A_\varepsilon \underbrace{R_\varepsilon}_{j-th} A_\varepsilon \cdots A_\varepsilon.$$

For h_0 as in (2.58) and $h > h_0$ we have

$$\|E_{h,\varepsilon}\| \leq \|R_\varepsilon\| \|A_\varepsilon\|^{2h_0} \|A_\varepsilon^{h_0}\|^{\lfloor h/h_0 \rfloor - 1} \leq \|R_\varepsilon\| \|A_\varepsilon\|^{2h_0} \|A_\varepsilon^{h_0}\|^{h/h_0 - 1},$$

where $\lfloor \cdot \rfloor$ denotes the integer part, and thus

$$\sum_{h=0}^{+\infty} \|E_{h,\varepsilon}\| \leq C \sum_{h=0}^{+\infty} \|A_\varepsilon^{h_0}\|^{h/h_0}$$

giving the absolute convergence of $\sum E_{h,\varepsilon}$. Summarizing we conclude that

$$(I - A_\varepsilon)^{-1} = \left(\frac{1}{2}I + K_\Omega\right)^{-1} + O(\varepsilon^{d-1}). \tag{2.59}$$

We lastly provide the proof the Theorem 2.5.1.

Proof of Theorem 2.5.1 To get (2.55), we need to analyze the two integrals J_1 and J_2 in (2.54).

For $\boldsymbol{x}, \boldsymbol{\zeta} \in \mathbb{R}^d$ with $\boldsymbol{x} \neq \boldsymbol{0}$ and ε sufficiently small, we consider the asymptotic expansion at first order of the fundamental solution, i.e.,

$$\Gamma(\boldsymbol{x} - \varepsilon\boldsymbol{\zeta}) = \Gamma(\boldsymbol{x}) - \varepsilon \nabla\Gamma(\boldsymbol{x}) \cdot \boldsymbol{\zeta} + O(\varepsilon^2).$$

Hence, for $\boldsymbol{x} \in \mathbb{R}^{d-1}$, we get

$$
\begin{aligned}
J_1 &= \varepsilon^{d-1} \int_{\partial\Omega} \Gamma(\boldsymbol{x} - \boldsymbol{z} - \varepsilon\boldsymbol{\zeta})\, g^\dagger(\boldsymbol{\zeta})\, d\sigma(\boldsymbol{\zeta}) \\
&= \varepsilon^{d-1} \Gamma(\boldsymbol{x} - \boldsymbol{z}) \int_{\partial\Omega} g^\dagger(\boldsymbol{\zeta})\, d\sigma(\boldsymbol{\zeta}) \\
&\quad - \varepsilon^{d} \nabla\Gamma(\boldsymbol{x} - \boldsymbol{z}) \cdot \int_{\partial\Omega} \boldsymbol{\zeta}\, g^\dagger(\boldsymbol{\zeta})\, d\sigma(\boldsymbol{\zeta}) + O(\varepsilon^{d+1}).
\end{aligned} \tag{2.60}
$$

For the second integral in (2.54), we find

$$
J_2 = -\varepsilon^{d-1} \int_{\partial\Omega} \frac{\partial}{\partial \boldsymbol{n}_\zeta} \Gamma(\boldsymbol{x} - \boldsymbol{z} - \varepsilon\boldsymbol{\zeta}) h^\dagger(\boldsymbol{\zeta}, \varepsilon)\, d\sigma(\boldsymbol{\zeta}),
$$

where the function $h^\dagger$ is given by

$$
h^\dagger(\boldsymbol{\zeta}, \varepsilon) = \left(\frac{1}{2} I + K_\varepsilon + \widetilde{D}_\varepsilon\right)^{-1} \left(S_\varepsilon g + \widetilde{S}_\varepsilon g\right)(\boldsymbol{z} + \varepsilon\boldsymbol{\zeta}). \tag{2.61}
$$

For $\boldsymbol{x}, \boldsymbol{\zeta} \in \mathbb{R}^d$ with $\boldsymbol{x} \neq \boldsymbol{0}$ and ε sufficiently small, it holds

$$
\nabla_{\boldsymbol{x}} \Gamma(\boldsymbol{x} + \varepsilon\boldsymbol{\zeta}) = \nabla_{\boldsymbol{x}} \Gamma(\boldsymbol{x}) + O(\varepsilon), \tag{2.62}
$$

therefore, taking advantage of the expansion (2.56),

$$
\begin{aligned}
J_2 &= \varepsilon^{d-1} \int_{\partial\Omega} \frac{\partial}{\partial \boldsymbol{n}_\zeta} \Gamma(\boldsymbol{x} - \boldsymbol{z}) h_\varepsilon^\dagger(\boldsymbol{\zeta})\, d\sigma(\boldsymbol{\zeta}) + O(\varepsilon^d) \\
&= \varepsilon^d \int_{\partial\Omega} \frac{\partial}{\partial \boldsymbol{n}_\zeta} \Gamma(\boldsymbol{x} - \boldsymbol{z}) \left(\tfrac{1}{2} I + K_\Omega\right)^{-1} S_\Omega g^\dagger(\boldsymbol{\zeta})\, d\sigma(\boldsymbol{\zeta}) + O(\varepsilon^{d+1}).
\end{aligned}
$$

The statement of the theorem follows putting together the results for J_1 and J_2. □

To close this section we want to show that the term $\left(\tfrac{1}{2} I + K_\Omega\right)^{-1} S_\Omega g(\boldsymbol{x})$, for $\boldsymbol{x} \in \partial\Omega$, which appears in the second integral in (2.55), can be interpreted as the solution of an exterior problem in the whole space. Specifically, this term is the trace of the solution on $\partial\Omega$ of this boundary value problem

$$
\begin{cases}
\Delta U = 0 & \text{in } \mathbb{R}^d \setminus \Omega \\
\dfrac{\partial U}{\partial \boldsymbol{n}} = g & \text{on } \partial\Omega \\
U \to 0 & \text{as } |\boldsymbol{x}| \to +\infty,
\end{cases} \tag{2.63}
$$

where the cavity Ω is such that $\boldsymbol{0} \in \Omega$.

Proposition 2.5.4 *Let us define* $h(\boldsymbol{x}) := U(\boldsymbol{x})\big|_{\boldsymbol{x}\in\partial\Omega}$, *then*

$$\left(\frac{1}{2}I + K_{\Omega}\right)^{-1} S_{\Omega} g(\boldsymbol{x}) = h(\boldsymbol{x}).$$

Proof The thesis comes from, as done in the proof of Theorem 2.3.4, by the application of the second Green's identity to the fundamental solution Γ and U in the domain $B_R(\mathbf{0}) \setminus \Omega$, with R sufficiently large. We define $D_{R,\epsilon} := B_R(\mathbf{0}) \setminus (\Omega \cup B_\epsilon(\boldsymbol{x}))$, with $\boldsymbol{x} \in (B_R(\mathbf{0}) \setminus \Omega)$ (see Fig. 2.3). By the second Green's identity, we get

$$\begin{aligned}
0 &= \int_{\partial D_{R,\epsilon}} \left[U(\boldsymbol{y})\frac{\partial}{\partial \boldsymbol{n}_{\boldsymbol{y}}}\Gamma(\boldsymbol{x}-\boldsymbol{y}) - \Gamma(\boldsymbol{x}-\boldsymbol{y})\frac{\partial}{\partial \boldsymbol{n}_{\boldsymbol{y}}}U(\boldsymbol{y})\right]d\sigma(\boldsymbol{y}) \\
&= \int_{\partial B_R(\mathbf{0})} \left[U(\boldsymbol{y})\frac{\partial}{\partial \boldsymbol{n}_{\boldsymbol{y}}}\Gamma(\boldsymbol{x}-\boldsymbol{y}) - \Gamma(\boldsymbol{x}-\boldsymbol{y})\frac{\partial}{\partial \boldsymbol{n}_{\boldsymbol{y}}}U(\boldsymbol{y})\right]d\sigma(\boldsymbol{y}) \\
&\quad - \int_{\partial B_\epsilon(\boldsymbol{x})} \left[U(\boldsymbol{y})\frac{\partial}{\partial \boldsymbol{n}_{\boldsymbol{y}}}\Gamma(\boldsymbol{x}-\boldsymbol{y}) - \Gamma(\boldsymbol{x}-\boldsymbol{y})\frac{\partial}{\partial \boldsymbol{n}_{\boldsymbol{y}}}U(\boldsymbol{y})\right]d\sigma(\boldsymbol{y}) \\
&\quad - \int_{\partial C} \left[U(\boldsymbol{y})\frac{\partial}{\partial \boldsymbol{n}_{\boldsymbol{y}}}\Gamma(\boldsymbol{x}-\boldsymbol{y}) - \Gamma(\boldsymbol{x}-\boldsymbol{y})\frac{\partial}{\partial \boldsymbol{n}_{\boldsymbol{y}}}U(\boldsymbol{y})\right]d\sigma(\boldsymbol{y}) \\
&:= I_1 - I_2 - \mathcal{I}(h, g)(\boldsymbol{x}).
\end{aligned}$$

Using the decay rate of harmonic functions in unbounded domains, see Theorem 2.2.2, the integral I_1 gives

$$|I_1| \le \left(\frac{C_1}{R^{2d-3}} + \frac{C_2}{R^{2d-3}}\right)\int_{\partial B_R(\mathbf{0})} d\sigma(\boldsymbol{y}) = \frac{C}{R^{d-2}},$$

where C is a positive constant. As $R \to +\infty$, I_1 tends to zero.

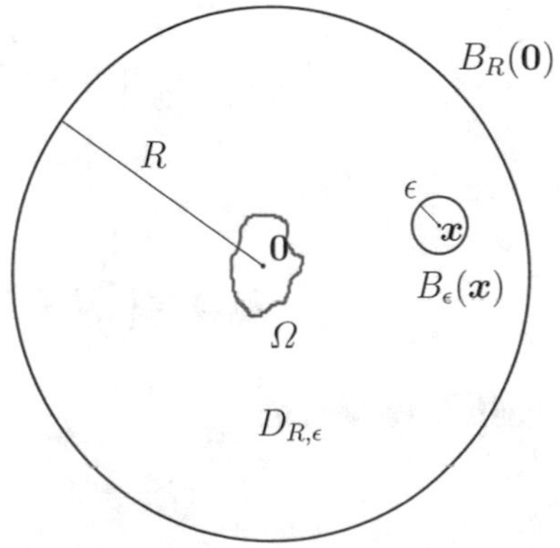

Fig. 2.3 Domain used to get the representation formula for U

Finally, we decompose I_2 as

$$I_2 = I_{21} - I_{22}$$
$$= \int_{\partial B_\epsilon(x)} U(y)\frac{\partial}{\partial \boldsymbol{n}_y}\Gamma(\boldsymbol{x}-\boldsymbol{y})\,d\sigma(\boldsymbol{y}) - \int_{\partial B_\epsilon(x)} \Gamma(\boldsymbol{x}-\boldsymbol{y})\frac{\partial}{\partial \boldsymbol{n}_y}U(\boldsymbol{y})\,d\sigma(\boldsymbol{y}).$$

Using the expression of Γ and the continuity of u, we derive

$$I_{21} = \int_{\partial B_\epsilon(x)} U(\boldsymbol{y})\frac{\partial}{\partial \boldsymbol{n}_y}\Gamma(\boldsymbol{x}-\boldsymbol{y})\,d\sigma(\boldsymbol{y}) = U(\boldsymbol{x})\int_{\partial B_\epsilon(x)} \frac{\partial}{\partial \boldsymbol{n}_y}\Gamma(\boldsymbol{x}-\boldsymbol{y})\,d\sigma(\boldsymbol{y})$$
$$+ \int_{\partial B_\epsilon(x)} (U(\boldsymbol{y}) - U(\boldsymbol{x}))\,\frac{\partial}{\partial \boldsymbol{n}_y}\Gamma(\boldsymbol{x}-\boldsymbol{y})d\sigma(\boldsymbol{y}),$$

which tends to $U(\boldsymbol{x})$ as $\varepsilon \to 0$. Moreover, it holds

$$|I_{22}| \le C' \sup_{y\in\partial B_\epsilon(x)} \left|\frac{\partial U(\boldsymbol{y})}{\partial \boldsymbol{n}_y}\right| \frac{1}{\epsilon^{d-2}} \int_{\partial B_\epsilon(x)} d\sigma(\boldsymbol{y}) = O(\epsilon)$$

which goes to zero as ϵ goes to zero.

In conclusion we have the following integral representation formula:

$$\begin{aligned} U(\boldsymbol{x}) &= -\mathcal{I}(h, g)(\boldsymbol{x}) \\ &= \int_{\partial\Omega} \left[\Gamma(\boldsymbol{x}-\boldsymbol{y})g(\boldsymbol{y}) - h(\boldsymbol{y})\frac{\partial}{\partial \boldsymbol{n}_y}\Gamma(\boldsymbol{x}-\boldsymbol{y})\right] d\sigma(\boldsymbol{y}) \\ &= S_\Omega g(\boldsymbol{x}) - D_\Omega h(\boldsymbol{x}), \qquad \boldsymbol{x} \in \mathbb{R}^d \setminus \overline{\Omega}, \end{aligned} \tag{2.64}$$

where h is the trace of U on the boundary of the cavity Ω. Therefore, on $\partial\Omega$ from single and double layer potentials properties

$$h(\boldsymbol{x}) = S_\Omega g(\boldsymbol{x}) - \left(-\frac{1}{2}I + K_\Omega\right) h(\boldsymbol{x}), \quad \boldsymbol{x} \in \partial\Omega,$$

hence

$$h(\boldsymbol{x}) = \left(\frac{1}{2}I + K_\Omega\right)^{-1} S_\Omega g(\boldsymbol{x}), \quad \boldsymbol{x} \in \partial\Omega,$$

that is the assertion. □

2.5.1 A Particular Neumann Condition

For the analogy with the elastic problem, in this section we consider a specific case of the Neumann boundary condition on the cavity C_ε. In particular, we get an explicit expression of the asymptotic expansion in terms of the polarization tensor and the fundamental solution.

Corollary 2.5.5 *Given $\boldsymbol{p} \in \mathbb{R}^d$, let the boundary datum in (2.27) given by*

$$g = -\boldsymbol{p} \cdot \boldsymbol{n}.$$

Then, the following expansion holds

$$u_\varepsilon(\boldsymbol{x}) = 2\,\varepsilon^d |\Omega| \nabla\Gamma(\boldsymbol{x} - \boldsymbol{z}) \cdot \mathbf{M}\boldsymbol{p} + O(\varepsilon^{d+1}), \qquad \boldsymbol{x} \in \mathbb{R}^{d-1}, \tag{2.65}$$

where $\mathbf{M}$ is the symmetric positive definite tensor given by

$$\mathbf{M} := \mathbf{I} + \frac{1}{|\Omega|} \int_{\partial\Omega} \big(\boldsymbol{n}_\zeta \otimes \boldsymbol{\Psi}(\boldsymbol{\zeta})\big)\, d\sigma(\boldsymbol{\zeta}), \tag{2.66}$$

and the auxiliary function $\boldsymbol{\Psi}$ has components Ψ_i, $i = 1, \ldots, d$, which solve

$$\begin{cases} \Delta\Psi_i = 0 & \text{in } \mathbb{R}^d \setminus \Omega \\ \dfrac{\partial\Psi_i}{\partial\boldsymbol{n}} = -n_i & \text{on } \partial\Omega \\ \Psi_i \to 0 & \text{as } |\boldsymbol{x}| \to +\infty. \end{cases} \tag{2.67}$$

Proof We define

$$I_1 := \nabla\Gamma(\boldsymbol{x} - \boldsymbol{z}) \cdot \int_{\partial\Omega} \boldsymbol{n}_\zeta \left(\frac{1}{2}I + K_\Omega\right)^{-1} S_\Omega[-\boldsymbol{p} \cdot \boldsymbol{n}](\boldsymbol{\zeta})\, d\sigma(\boldsymbol{\zeta}),$$

$$I_2 := \nabla\Gamma(\boldsymbol{x} - \boldsymbol{z}) \cdot \int_{\partial\Omega} \boldsymbol{\zeta}\, \boldsymbol{p} \cdot \boldsymbol{n}_\zeta\, d\sigma(\boldsymbol{\zeta}).$$

Then, the expansion (2.55) with $g = -\boldsymbol{p} \cdot \boldsymbol{n}$ gives

$$\begin{aligned} \frac{1}{2} u_\varepsilon(\boldsymbol{x}) &= -\varepsilon^{d-1}\Gamma(\boldsymbol{x} - \boldsymbol{z}) \int_{\partial\Omega} \boldsymbol{p} \cdot \boldsymbol{n}_\zeta\, d\sigma(\boldsymbol{\zeta}) + I_1 + I_2 + O(\varepsilon^{d+1}) \\ &= I_1 + I_2 + O(\varepsilon^{d+1}), \end{aligned} \tag{2.68}$$

where the first term in the asymptotic expansion is null thanks to the divergence theorem. From Eq. (2.63), with $g = -\boldsymbol{p} \cdot \boldsymbol{n}$, since the problem for U is linear, we can decompose U as $U = \sum_i U_i$ where the functions U_i, for $i = 1, \cdots, d$, solve

$$\begin{cases} \Delta U_i = 0 & \text{in } \mathbb{R}^d \setminus \Omega \\ \dfrac{\partial U_i}{\partial \boldsymbol{n}} = -p_i n_i & \text{on } \partial\Omega \\ U_i \to 0 & \text{as } |\boldsymbol{x}| \to +\infty. \end{cases}$$

According to the definition of the functions Ψ_i, we have that $U = \boldsymbol{p} \cdot \boldsymbol{\Psi}$. Therefore, utilizing the result of Proposition 2.5.4, the term I_1 is also equal to

$$\begin{aligned} I_1 &= \nabla\Gamma(\boldsymbol{x} - \boldsymbol{z}) \cdot \int_{\partial\Omega} (\boldsymbol{\Psi}(\boldsymbol{\zeta}) \cdot \boldsymbol{p})\, \boldsymbol{n}_\zeta \, d\sigma(\boldsymbol{\zeta}) \\ &= \nabla\Gamma(\boldsymbol{x} - \boldsymbol{z}) \cdot \int_{\partial\Omega} \big(\boldsymbol{n}_\zeta \otimes \boldsymbol{\Psi}(\boldsymbol{\zeta})\big)\, \boldsymbol{p} \, d\sigma(\boldsymbol{\zeta}). \end{aligned}$$

To deal with the term I_2, we first observe that

$$\int_{\partial\Omega} \big(\boldsymbol{n}_\zeta \otimes \boldsymbol{\zeta}\big)\, d\sigma(\boldsymbol{\zeta}) = |\Omega| \mathbf{I}.$$

Indeed, for $\boldsymbol{n}_\zeta = (n_{\zeta,1}, \ldots, n_{\zeta,d})$, for any $i, j \in \{1, \ldots, d\}$, it follows

$$\begin{aligned} \int_{\partial\Omega} \zeta_i\, n_{\zeta,j}\, d\sigma(\boldsymbol{\zeta}) &= \int_{\partial\Omega} \boldsymbol{n}_\zeta \cdot \zeta_i \boldsymbol{e}_j \, d\sigma(\boldsymbol{\zeta}) \\ &= \int_{\Omega} \operatorname{div}\big(\zeta_i \boldsymbol{e}_j\big)\, d\boldsymbol{\zeta} = \int_{\Omega} \boldsymbol{e}_j \cdot \boldsymbol{e}_i \, d\boldsymbol{\zeta} = |\Omega| \delta_{ij}, \end{aligned}$$

where $\boldsymbol{e}_j$ is the j-th unit vector of $\mathbb{R}^d$. Hence, it follows

$$I_2 = \nabla\Gamma(\boldsymbol{x} - \boldsymbol{z}) \cdot \int_{\partial\Omega} (\boldsymbol{\zeta} \otimes \boldsymbol{n}_\zeta)\, \boldsymbol{p}\, d\sigma(\boldsymbol{\zeta}) = |\Omega| \nabla\Gamma(\boldsymbol{x} - \boldsymbol{z}) \cdot \boldsymbol{p}.$$

Gathering the expressions of I_1 and I_2, we obtain formula (2.65).

Next we show the properties of the tensor $\mathbf{M}$ defined in (2.66). Indeed, the symmetry follows from

$$\begin{aligned} \int_{\partial\Omega} \Psi_i(\boldsymbol{\zeta})\, n_{\zeta,j}\, d\sigma(\boldsymbol{\zeta}) &= -\int_{\partial\Omega} \Psi_i(\boldsymbol{\zeta}) \frac{\partial \Psi_j}{\partial \boldsymbol{n}}(\boldsymbol{\zeta})\, d\sigma(\boldsymbol{\zeta}) \\ &= \int_{\mathbb{R}^d \setminus \Omega} \operatorname{div}\big(\Psi_i(\boldsymbol{\zeta}) \nabla\Psi_j(\boldsymbol{\zeta})\big)\, d\boldsymbol{\zeta} \\ &= \int_{\mathbb{R}^d \setminus \Omega} \nabla\Psi_i(\boldsymbol{\zeta}) \cdot \nabla\Psi_j(\boldsymbol{\zeta})\, d\boldsymbol{\zeta}, \end{aligned}$$

where the last term is obviously symmetric. Taking $\boldsymbol{\eta} \in \mathbb{R}^d$, we consider

$$\boldsymbol{\eta} \cdot \mathbf{M}\boldsymbol{\eta} = |\boldsymbol{\eta}|^2 + \frac{1}{|\Omega|} \int_{\partial\Omega} (\boldsymbol{n}_{\zeta} \cdot \boldsymbol{\eta})(\Psi(\boldsymbol{\zeta}) \cdot \boldsymbol{\eta})\, d\sigma(\boldsymbol{\zeta}).$$

The positivity of the tensor follows from the divergence theorem, integration by parts and the definition of the function $\boldsymbol{\Psi}$, in fact

$$\begin{aligned}
\int_{\partial\Omega} (\boldsymbol{n}_{\zeta} \cdot \boldsymbol{\eta})(\boldsymbol{\Psi}(\boldsymbol{\zeta}) \cdot \boldsymbol{\eta})\, d\sigma(\boldsymbol{\zeta}) &= -\int_{\partial\Omega} \frac{\partial}{\partial \boldsymbol{n}} (\boldsymbol{\Psi}(\boldsymbol{\zeta}) \cdot \boldsymbol{\eta})(\boldsymbol{\Psi}(\boldsymbol{\zeta}) \cdot \boldsymbol{\eta})\, d\sigma(\boldsymbol{\zeta}) \\
&= \int_{\mathbb{R}^d \setminus \Omega} \operatorname{div}\left((\boldsymbol{\Psi}(\boldsymbol{\zeta}) \cdot \boldsymbol{\eta}) \nabla (\boldsymbol{\Psi}(\boldsymbol{\zeta}) \cdot \boldsymbol{\eta})\right) d\sigma(\boldsymbol{\zeta}) \\
&= \int_{\mathbb{R}^d \setminus \Omega} \left|\nabla (\boldsymbol{\Psi}(\boldsymbol{\zeta}) \cdot \boldsymbol{\eta})\right|^2 d\boldsymbol{\zeta},
\end{aligned}$$

hence $\boldsymbol{\eta} \cdot \mathbf{M}\boldsymbol{\eta} > 0$. □

We close this chapter showing that in the case of a spherical cavity, i.e., when Ω is a sphere, the auxiliary function $\boldsymbol{\Psi}$ can be determined explicitly. From this, it follows that we also have an explicit formula for the polarization tensor $\mathbf{M}$. If $\Omega = \{\boldsymbol{x} \in \mathbb{R}^3 : |\boldsymbol{x}| < 1\}$, then a direct calculation through spherical coordinates shows that, for $i = 1, 2, 3$, it holds $\Psi_i(\boldsymbol{x}) = x_i/(2|\boldsymbol{x}|^3)$, and thus

$$\Psi_i(\boldsymbol{\zeta}) = \frac{1}{2} \zeta_i, \qquad \boldsymbol{\zeta} \in \partial\Omega.$$

As a consequence, the polarization tensor is a multiple of the identity and, precisely,

$$\mathbf{M} = \frac{3}{2} |\Omega| \mathbf{I} = 2\pi \mathbf{I}.$$

Then, the asymptotic expansion (2.65) becomes

$$u_{\varepsilon}(\boldsymbol{x}) = 4\pi \varepsilon^3 \nabla \Gamma(\boldsymbol{x} - \boldsymbol{z}) \cdot \boldsymbol{p} + O(\varepsilon^4), \qquad \boldsymbol{x} \in \mathbb{R}^2.$$

Explicit formulas can be provided also in the case of ellipsoidal cavities (see [18, 21, 22]).

Chapter 3
Analysis of the Elastic Model

In this chapter we study the linear elastic model presented in the introduction. We recall that it is applied in volcanology to describe the surface deformation effects caused by a magma chamber embedded into Earth's interior and exerting on it a uniform hydrostatic pressure. From a mathematical point of view, the modeling assumptions translates into a Neumann boundary value problem for the classic Lamé system in a half-space with an embedded pressurized cavity. To be more precise, the boundary conditions are traction-free for the air/crust boundary and uniformly hydrostatic for the chamber boundary.

Therefore, representing with $\boldsymbol{u}$ the displacement vector field, we need to study the following boundary value problem:

$$\begin{cases} \operatorname{div}(\mathbb{C}\widehat{\nabla}\boldsymbol{u}) = \boldsymbol{0} & \text{in } \mathbb{R}^3_- \setminus C \\ \dfrac{\partial \boldsymbol{u}}{\partial \boldsymbol{\nu}} = p\,\boldsymbol{n} & \text{on } \partial C \\ \dfrac{\partial \boldsymbol{u}}{\partial \boldsymbol{\nu}} = \boldsymbol{0} & \text{on } \mathbb{R}^2, \end{cases} \tag{3.1}$$

where $\mathbb{C}$ is the elasticity tensor, C is the cavity, p is a constant representing the pressure, and $\widehat{\nabla}\boldsymbol{u} = \frac{1}{2}\big(\nabla\boldsymbol{u} + \nabla\boldsymbol{u}^T\big)$ the strain tensor. With $\partial\boldsymbol{u}/\partial\boldsymbol{\nu}$ we depict the conormal derivative on the boundary of a domain; that is, the traction vector $\partial\boldsymbol{u}/\partial\boldsymbol{\nu} := (\mathbb{C}\widehat{\nabla}\boldsymbol{u})\boldsymbol{n}$.

As done in the previous chapter for the scalar model, we provide two different ways to prove the well-posedness of the problem: one is to add to (3.1) the behavior at infinity of $\boldsymbol{u}$ and its gradient, i.e., $\boldsymbol{u} = o(\boldsymbol{1})$ and $\nabla\boldsymbol{u} = o(|\boldsymbol{x}|^{-1})$ as $|\boldsymbol{x}| \to \infty$. Then, we utilize an appropriate integral formulation of (3.1) so to reduce the well-posedness to the study of the invertibility of some integral operators.

Another way is to prove the well-posedness of (3.1) through a weak formulation based on weighted Sobolev spaces, as explained in the previous chapter. Then, the well-posedness follows by the application of the Lax–Milgram theorem. Using this

A. Aspri, *An Elastic Model for Volcanology*, Lecture Notes in Geosystems Mathematics and Computing, https://doi.org/10.1007/978-3-030-31475-0_3

approach we naturally get stability estimates of the solution in the specific weighted Sobolev space.

After the well-posedness, we first analyze the inverse problem of determining the pressurized cavity from partial measurements of the displacement field on the boundary of the half-space proving uniqueness and stability estimates. To this end, we have to prove a quantitative weighted Poincaré-type inequality and a Korn-type inequality in $\mathbb{R}^3_- \setminus C$ since, from the Lax–Milgram theorem, we need quantitative stability estimates of the solution of the direct problem. Then, using the integral representation formula utilized to prove the well-posedness, we provide an asymptotic expansion of the solution of (3.1) when the cavity C is small compared with the distance from the boundary of the half-space $\mathbb{R}^2$. Specifically, assuming that the chamber is centered at some fixed point z and has diameter $r > 0$ and depth d, we derive rigorously the principal term in the asymptotic expansion for the surface deformation as $\varepsilon = r/d \to 0^+$. Such formula provides a rigorous proof of the Mogi point source model in the case of spherical cavities, presented in Chap. 1, generalizing it to the case of cavities of arbitrary shape. For the application that we have in mind, we focus the attention only to the dimensional case $d = 3$.

This chapter is organized as follows: In Sect. 3.1 we recall some arguments about linear elasticity, layer potentials techniques, and the Neumann function for the Lamé operator in the half-space. In Sect. 3.2 we analyze the well-posedness of the direct problem both via an integral representation formula and via a weak formulation. Core of the chapter are the last two sections where in Sect. 3.3 we analyze the inverse problem of detecting the cavity through boundary measurements while in Sect. 3.4 we derive the asymptotic formula for the boundary displacement field when the cavity C is small compared to the distance from the boundary of the domain. As a consequence of the asymptotic expansion, we retrieve the classical Mogi's formula.

3.1 Lamé Operator and Layer Potentials

Let Ω be a bounded Lipschitz domain in $\mathbb{R}^3$ representing the region occupied by a homogeneous and isotropic elastic medium. Let λ and μ be the Lamé parameters, i.e., the compression modulus and the shear modulus, we define the fourth-order elasticity tensor

$$\mathbb{C} := \lambda \mathbf{I} \otimes \mathbf{I} + 2\mu \mathbb{I},$$

where $\mathbf{I}$ is the identity matrix and $\mathbb{I}$ is the identity fourth-order tensor such that $\mathbb{I}\mathbf{A} = \widehat{\mathbf{A}}$. We assume that $\mathbb{C}$ satisfies the minor and major symmetry conditions, i.e.,

$$\mathbb{C}_{ijkh} = \mathbb{C}_{khij} = \mathbb{C}_{jikh},$$

for all $i, j, k, h = 1, 2, 3$. If λ and μ satisfies the physical range

$$3\lambda + 2\mu > 0, \qquad \text{and} \qquad \mu > 0, \tag{3.2}$$

the elasticity tensor $\mathbb{C}$ is said to be strongly convex, i.e.,

$$\mathbb{C}\widehat{\mathbf{A}} : \widehat{\mathbf{A}} \geq \eta_0 |\widehat{\mathbf{A}}|^2, \tag{3.3}$$

where $\eta_0 = \min\{2\mu, 2\mu + 3\lambda\}$. For a more in-depth discussion see [62, 87].

In linear elasticity, it is also common to use the Poisson ratio ν which is related to λ and μ by the identity $\nu = \lambda/2(\lambda + \mu)$.

3.1.1 Some Preliminary Notions

In a homogeneous and isotropic elastic medium, the elastostatic Lamé operator $\mathcal{L}$ is defined by

$$\mathcal{L}\boldsymbol{u} := \operatorname{div}(\mathbb{C}\widehat{\nabla}\boldsymbol{u}) = \mu\Delta\boldsymbol{u} + (\lambda + \mu)\nabla\operatorname{div}\boldsymbol{u},$$

where $\boldsymbol{u}$ represents the vector of the displacements. In terms of the Poisson ratio it becomes $\mathcal{L}\boldsymbol{u} = \mu(\Delta\boldsymbol{u} + 1/(1 - 2\nu)\nabla\operatorname{div}\boldsymbol{u})$.

The explicit expression of the conormal derivative is given by

$$\frac{\partial \boldsymbol{u}}{\partial \boldsymbol{\nu}} := (\mathbb{C}\widehat{\nabla}\boldsymbol{u})\boldsymbol{n} = \lambda(\operatorname{div}\boldsymbol{u})\boldsymbol{n} + 2\mu(\widehat{\nabla}\boldsymbol{u})\boldsymbol{n},$$

or, equivalently,

$$\frac{\partial \boldsymbol{u}}{\partial \boldsymbol{\nu}} = 2\mu\frac{\partial \boldsymbol{u}}{\partial \boldsymbol{n}} + \lambda(\operatorname{div}\boldsymbol{u})\boldsymbol{n} + \mu(\boldsymbol{n} \times \operatorname{rot}\boldsymbol{u}).$$

In this chapter we use both the notation $\frac{\partial \boldsymbol{u}}{\partial \boldsymbol{\nu}}$ and $(\mathbb{C}\widehat{\nabla}\boldsymbol{u})\boldsymbol{n}$ to indicate the conormal derivative. In particular, we use the first one to simplify the notation in some cases.

For the sequel, we recall that the strong convexity of the tensor $\mathbb{C}$ implies the strong ellipticity of the Lamé operator which corresponds to the request $\mu > 0$ and $\lambda + 2\mu > 0$, see [87].

We recall the Green's formulas for the Lamé system which are called Betti's formulas. They can be obtained by integration by parts, see, for example, [27, 104]. Given a bounded Lipschitz domain $D \subset \mathbb{R}^3$ and two vectors $\boldsymbol{u},\boldsymbol{v} \in \mathbb{R}^3$, the *first Betti formula* is

$$\int_{\partial D} \boldsymbol{u} \cdot \frac{\partial \boldsymbol{v}}{\partial \boldsymbol{\nu}}\, d\sigma(\boldsymbol{x}) = \int_D \boldsymbol{u} \cdot \mathcal{L}\boldsymbol{v}\, d\boldsymbol{x} + \int_D \mathbb{C}\widehat{\nabla}\boldsymbol{u} : \widehat{\nabla}\boldsymbol{v}\, d\boldsymbol{x}, \tag{3.4}$$

where the quadratic form $Q := \mathbb{C}\widehat{\nabla}\boldsymbol{u} : \widehat{\nabla}\boldsymbol{v}$ is also equal to

$$Q(\boldsymbol{u}, \boldsymbol{v}) = \lambda(\operatorname{div} \boldsymbol{u})(\operatorname{div} \boldsymbol{v}) + 2\mu\widehat{\nabla}\boldsymbol{u} : \widehat{\nabla}\boldsymbol{v}.$$

From (3.4) it is straightforward to find the *second Betti formula*

$$\int_D (\boldsymbol{u} \cdot \mathcal{L}\boldsymbol{v} - \boldsymbol{v} \cdot \mathcal{L}\boldsymbol{u})\, dx = \int_{\partial D} \left(\boldsymbol{u} \cdot \frac{\partial \boldsymbol{v}}{\partial \boldsymbol{\nu}} - \boldsymbol{v} \cdot \frac{\partial \boldsymbol{u}}{\partial \boldsymbol{\nu}} \right) d\sigma(\boldsymbol{x}). \tag{3.5}$$

These formulas will be used to prove the well-posedness of the (3.1), both using an integral and a weighted Sobolev space approach, and to get an integral representation formula for it. For this purpose, we need to introduce the fundamental solution of the Lamé operator: the *Kelvin matrix* $\boldsymbol{\Gamma}$ (or *Kelvin-Somigliana matrix*), i.e., the solution to the system

$$\operatorname{div}(\mathbb{C}\widehat{\nabla}\boldsymbol{\Gamma}) = \delta_{\boldsymbol{0}}\mathbf{I}, \qquad \boldsymbol{x} \in \mathbb{R}^3 \setminus \{\boldsymbol{0}\},$$

where $\delta_{\boldsymbol{0}}$ is the Dirac function centered at $\boldsymbol{0}$. Setting $C_{\mu,\nu} := 1/\{16\pi\mu(1-\nu)\}$, the explicit expression of $\boldsymbol{\Gamma} = (\Gamma_{ij})$ is

$$\Gamma_{ij}(\boldsymbol{x}) = -C_{\mu,\nu}\left\{ \frac{(3-4\nu)\delta_{ij}}{|\boldsymbol{x}|} + \frac{x_i x_j}{|\boldsymbol{x}|^3} \right\}, \qquad i, j = 1, 2, 3, \tag{3.6}$$

where δ_{ij} is the Kronecker symbol and Γ_{ij} stands for the i-th component of the displacement when a force is applied in the j-th direction at the point $\boldsymbol{0}$. We recall that the gradient of $\boldsymbol{\Gamma}$ is given by

$$\frac{\partial \Gamma_{ij}}{\partial x_k}(\boldsymbol{x}) = C_{\mu,\nu}\left\{ \frac{(3-4\nu)\delta_{ij}x_k - \delta_{ik}x_j - \delta_{jk}x_i}{|\boldsymbol{x}|^3} + \frac{3x_i x_j x_k}{|\boldsymbol{x}|^5} \right\}, \qquad i, j, k = 1, 2, 3. \tag{3.7}$$

Therefore from (3.6) and (3.7) it is straightforward to infer the following behavior at infinity, i.e.,

$$|\boldsymbol{\Gamma}(\boldsymbol{x})| = O(|\boldsymbol{x}|^{-1}) \quad \text{and} \quad |\nabla\boldsymbol{\Gamma}(\boldsymbol{x})| = O(|\boldsymbol{x}|^{-2}) \qquad \text{as} \quad |\boldsymbol{x}| \to \infty. \tag{3.8}$$

3.1.2 Layer Potentials for the Lamé Operator

In this section, we recall the main properties of the single and double layer potentials connected to the Lamé system $\mathcal{L}$. Given $\boldsymbol{\varphi} \in \boldsymbol{L}^2(\partial\Omega)$ (see [21, 27, 104]), the *single* and *double* layer potentials are defined by

$$\begin{aligned}
\mathbf{S}^\Gamma\varphi(x) &:= \int_{\partial\Omega} \mathbf{\Gamma}(x-y)\varphi(y)\,d\sigma(y), && x\in\mathbb{R}^3,\\
\mathbf{\Omega}^\Gamma\varphi(x) &:= \int_{\partial\Omega} \frac{\partial\boldsymbol{\Gamma}}{\partial\boldsymbol{\nu}(y)}(x-y)\varphi(y)\,d\sigma(y), && x\in\mathbb{R}^3\setminus\partial\Omega,
\end{aligned}\tag{3.9}$$

where $\partial\mathbf{\Gamma}/\partial\boldsymbol{\nu}$ denotes the conormal derivative applied to each column of the matrix $\mathbf{\Gamma}$.

We summarize here some properties of these operators:

(i) By definition, $\mathbf{S}^\Gamma\varphi(x)$ and $\mathbf{D}^\Gamma\varphi(x)$ satisfy the Lamé system in $\mathbb{R}^3\setminus\partial\Omega$.
(ii) $\mathbf{S}^\Gamma\varphi(x) = O(|x|^{-1})$ and $\mathbf{D}^\Gamma\varphi(x) = O(|x|^{-2})$ as $|x|\to+\infty$.

Next, we introduce $\mathbf{K}$ and $\mathbf{K}^*$ that is the L^2-adjoint Neumann-Poincaré boundary integral operators defined, in the sense of Cauchy principal value, by

$$\begin{aligned}
\mathbf{K}\varphi(x) &:= \text{p.v.}\int_{\partial\Omega} \frac{\partial\mathbf{\Gamma}}{\partial\boldsymbol{\nu}(y)}(x-y)\varphi(y)\,d\sigma(y),\\
\mathbf{K}^*\varphi(x) &:= \text{p.v.}\int_{\partial\Omega} \frac{\partial\mathbf{\Gamma}}{\partial\boldsymbol{\nu}(x)}(x-y)\varphi(y)\,d\sigma(y).
\end{aligned}$$

As in the previous chapter, in the sequel the subscripts $+$ and $-$ indicate the limits from outside and inside of the set Ω, respectively (see (2.17) for the definition). We recall that $t_1,\cdots,t_{d-1}$ represent an orthonormal basis for the tangent plane to $\partial\Omega$ and $\partial/\partial t = \sum_{k=1}^{d-1}\partial/\partial t_k\, t_k$ is the tangential derivative on $\partial\Omega$.

The following theorem about the jump relations of single and double potentials for Lipschitz domains is due to Dahlberg et al. [66].

Theorem 3.1.1 ([66]) *Let Ω be a bounded Lipschitz domain in $\mathbb{R}^3$. For $\varphi\in L^2(\partial\Omega)$, the following relations hold, a.e., on $\partial\Omega$,*

$$\begin{aligned}
\mathbf{D}^\Gamma\varphi\Big|_\pm(x) &= \left(\mp\tfrac{1}{2}\mathbf{I}+\mathbf{K}\right)\varphi(x),\\
\frac{\partial\mathbf{S}^\Gamma\varphi}{\partial\boldsymbol{\nu}}\Big|_\pm(x) &= \left(\pm\tfrac{1}{2}\mathbf{I}+\mathbf{K}^*\right)\varphi(x),\\
\frac{\partial\mathbf{S}^\Gamma\varphi}{\partial t}\Big|_-(x) &= \frac{\partial\mathbf{S}^\Gamma\varphi}{\partial t}\Big|_+(x).
\end{aligned}\tag{3.10}$$

It is worth noticing that the two operators $\mathbf{K}$ and $\mathbf{K}^*$ are not compact even on smooth domains, in contrast with the analogous operators for the Laplace equation (see [21] and the considerations in the previous chapter), due to the presence in their kernels of the terms

$$\frac{n_i(x_j - y_j)}{|\boldsymbol{x} - \boldsymbol{y}|^3} - \frac{n_j(x_i - y_i)}{|\boldsymbol{x} - \boldsymbol{y}|^3}, \qquad i \neq j, \tag{3.11}$$

which make the kernel not integrable. Indeed, even in the case of smooth domains, we cannot approximate locally the terms $\boldsymbol{n} \times (\boldsymbol{x} - \boldsymbol{y})$ with a smooth function, that is in terms of powers of $|\boldsymbol{x} - \boldsymbol{y}|$ via Taylor expansion, in order to obtain an integrable kernel on $\partial\Omega$. Therefore, the analysis to prove invertibility of the operators in (3.10) is complicated and usually based on a regularization procedure (see [104]) in the case of smooth domains. For Lipschitz domains the analysis is much more involved and, as for the Laplace operator, based on Rellich formulas. These results are contained in [66] and its companion article [79]. We recall here only the main aspects for the three-dimensional case.

Let Ψ be the vector space of all linear solutions of the equations

$$\begin{cases} \operatorname{div}(\mathbb{C}\widehat{\nabla}\boldsymbol{\psi}) = 0, & \text{in } \Omega \\ \dfrac{\partial \boldsymbol{\psi}}{\partial \boldsymbol{\nu}} = 0 & \text{on } \partial\Omega, \end{cases}$$

or, alternatively

$$\Psi := \left\{ \boldsymbol{\psi} : \nabla\boldsymbol{\psi} + (\nabla\boldsymbol{\psi})^T = 0 \right\}.$$

This space is independent from the Lamé parameters and its dimension is 6. A function $\boldsymbol{\psi} \in \Psi$ is called infinitesimal rigid motion. We recall that $\boldsymbol{\psi}$ can be expressed as

$$\boldsymbol{\psi} = \boldsymbol{a} + \mathbf{A}\boldsymbol{x}, \tag{3.12}$$

where $\mathbf{A}$ is a skew-symmetric matrix and $\boldsymbol{a} \in \mathbb{R}^3$. We define

$$\boldsymbol{L}^2_\Psi(\partial\Omega) := \left\{ \boldsymbol{f} \in \boldsymbol{L}^2(\partial\Omega) : \int_{\partial\Omega} \boldsymbol{f} \cdot \boldsymbol{w}\, d\sigma = 0, \forall \boldsymbol{w} \in \Psi \right\}.$$

We have

Proposition 3.1.2 ([66]) *The operators*

$$-\frac{1}{2}\mathbf{I} + \mathbf{K}^* : \boldsymbol{L}^2_\Psi(\partial\Omega) \to \boldsymbol{L}^2_\Psi(\partial\Omega)$$

$$\frac{1}{2}\mathbf{I} + \mathbf{K}^* : \boldsymbol{L}^2(\partial\Omega) \to \boldsymbol{L}^2(\partial\Omega)$$

are injective.

We omit the proof since is similar to that one of the scalar case.

The range of $-1/2\mathbf{I} + \mathbf{K}^*$ as an operator on all $L^2(\partial\Omega)$ is contained in $L^2_\Psi(\partial\Omega)$ since

$$\int_{\partial\Omega} \frac{\partial \mathbf{S}^\Gamma \boldsymbol{\varphi}}{\partial \boldsymbol{\nu}}\Big|_+ (\boldsymbol{x})\boldsymbol{\psi}(\boldsymbol{x})\, d\sigma = \int_{\partial\Omega} \mathbf{S}^\Gamma \boldsymbol{\varphi} \frac{\partial \boldsymbol{\psi}}{\partial \boldsymbol{\nu}}\, d\sigma(\boldsymbol{x}) = 0,$$

for all $\boldsymbol{\psi} \in \Psi$. This is because $\boldsymbol{\psi}$ is a solution to the elastostatic systems satisfying $\partial\boldsymbol{\psi}/\partial\boldsymbol{\nu} = 0$.

In addition, it holds

Proposition 3.1.3 ([66]) *The operators*

$$-\frac{1}{2}\mathbf{I} + \mathbf{K}^* : L^2_\Psi(\partial\Omega) \to L^2_\Psi(\partial\Omega)$$
$$\frac{1}{2}\mathbf{I} + \mathbf{K}^* : L^2(\partial\Omega) \to L^2(\partial\Omega)$$

have closed range.

The key point to show that these two boundary operators have closed range, as in the case of the Laplace operator, is the following inequality:

$$C^{-1}\Big\|\Big(-\frac{1}{2}\mathbf{I}+\mathbf{K}^*\Big)\boldsymbol{\varphi}\Big\|_{L^2(\partial\Omega)} \le \Big\|\Big(\frac{1}{2}\mathbf{I}+\mathbf{K}^*\Big)\boldsymbol{\varphi}\Big\|_{L^2(\partial\Omega)} \le C\Big\|\Big(-\frac{1}{2}\mathbf{I}+\mathbf{K}^*\Big)\boldsymbol{\varphi}\Big\|_{L^2(\partial\Omega)},$$

where C is a constant independent of $\boldsymbol{\varphi} \in L^2(\partial\Omega)$. However, we stress that the analysis to get the equivalence of the norms in the elastic case is really complicated since it is based on the twine of Rellich formulas for the Lamé operators, estimates derived from them, Korn inequalities and results on the biharmonic equations.

In order to prove the invertibility of the operators, it remains to show dense range. To do that one can make use of the result on the invertibility for the same operators in the case of smooth domains. The minimum regularity we request on the domain is, at least, C^1 but here, without loss of generality, we consider C^∞ domains. As stated before, even if we use smooth domains we cannot apply the Fredholm theory because $\mathbf{K}$ and $\mathbf{K}^*$ are not compact operators. However, the difference $\mathbf{K}-\mathbf{K}^*$ yields a compact operator, see [66] for details.

The following proposition is needed.

Proposition 3.1.4 *Let H be a Hilbert space. If $T : H \to H$ is a bounded linear operator with closed range, with null space of dimension $l < \infty$, and such that $T - T^*$ is compact, then the range of T has codimension l also.*

Now, we state the invertibility result for smooth domains.

Lemma 3.1.5 ([66]) *Let Ω be a bounded smooth domain with connected boundary in $\mathbb{R}^3$. Let us consider the operators $\pm 1/2\mathbf{I} + \mathbf{K}^*$ on $\partial\Omega$. Then*

(i) $-\frac{1}{2}\mathbf{I}+\mathbf{K}^* : L^2_\Psi(\partial\Omega) \to L^2_\Psi(\partial\Omega)$
(ii) $\frac{1}{2}\mathbf{I}+\mathbf{K}^* : L^2(\partial\Omega) \to L^2(\partial\Omega)$

are invertible operators.

Proof Let us prove (i) (the same argument yield for (ii) also). From the previous two propositions we know that the operator $-1/2\mathbf{I}+\mathbf{K}^*$ is one-to-one and has closed range. Moreover, the dimension of the null space is less than or equal to 6 and the codimension is greater than or equal to 6. Since $(-1/2\mathbf{I}+\mathbf{K}^*)-(-1/2\mathbf{I}+\mathbf{K})$ is compact, applying the Proposition 3.1.4, we have the assertion. □

Now, we briefly explain the sequence of steps to deduce the invertibility of the operators $\pm 1/2\mathbf{I}+\mathbf{K}^*$ in the case of Lipschitz domains, giving only the main ideas.

The starting point is to consider a sequence of C^∞ domains, which we call Ω_j, that converge to the Lipschitz domain Ω (for all the details see Theorem 1.12 in [131]). In such a scheme the $\partial\Omega_j$ can be projected homeomorphically to $\partial\Omega$ so that the boundaries converge uniformly and so that the Lipschitz characters of the Ω_j are controlled by that of Ω. In fact, the unit normal vectors to the Ω_j will converge pointwise, a.e., to those for Ω and in $L^p(\partial\Omega)$ for all $1 \le p \le \infty$. If $\mathbf{K}_j$ denotes the singular operators defined on $\partial\Omega_j$ we may project it onto $\partial\Omega$ and prove that

$$\lim_{j\to+\infty}\left\|\mathbf{K}_j^* f-\mathbf{K}^* f\right\|_{L^2(\partial\Omega_j)}=0,$$

and a result analogous for the adjoint operator $\mathbf{K}$. Then, since

$$\begin{aligned}\dim\Big(\mathrm{Ker}\Big(\pm\frac{1}{2}\mathbf{I}+\mathbf{K}^*\Big)\Big) &\le \dim\Big(\mathrm{Ker}\Big(\pm\frac{1}{2}\mathbf{I}+\mathbf{K}_j^*\Big)\Big)\\ &=\dim\Big(\mathrm{Coker}\Big(\pm\frac{1}{2}\mathbf{I}+\mathbf{K}_j^*\Big)\Big)=l\\ &\le \dim\Big(\mathrm{Coker}\Big(\pm\frac{1}{2}\mathbf{I}+\mathbf{K}^*\Big)\Big),\end{aligned}$$

where $l<\infty$ is independent of j, under other suitable assumptions, it can be proven that

$$\dim\Big(\mathrm{Coker}\Big(\pm\frac{1}{2}\mathbf{I}+\mathbf{K}^*\Big)\Big)=\dim\Big(\mathrm{Ker}\Big(\pm\frac{1}{2}\mathbf{I}+\mathbf{K}^*\Big)\Big)=l.$$

Finally, using the invertibility Lemma 3.1.5 about smooth domains we find

Theorem 3.1.6 *Let Ω be a bounded Lipschitz domain with connected boundary in $\mathbb{R}^3$. Then*

(i) $-\frac{1}{2}\mathbf{I}+\mathbf{K}^* :: L^2_\Psi(\partial\Omega) \to L^2_\Psi(\partial\Omega)$
(ii) $\frac{1}{2}\mathbf{I}+\mathbf{K}^* : L^2(\partial\Omega) \to L^2(\partial\Omega)$

are invertible operators.

3.1.3 Neumann Function in the Half-Space

In this section we find the retrieve expression of the Neumann function for the half-space which will be of help to find the integral representation formula of a solution of (3.1). It was presented by Mindlin in [112] for the first time and secondly in [113]. He used two different techniques to find its explicit expression. In [112] he used Galerkin vectors and nuclei of strains of the theory of linear elasticity; in [113] he utilized the Papkovich–Neuber representations of a displacement vector field and the potential theory. Here, we recall this second approach since it is based on mathematical techniques. These results are contained in [36].

Let us consider the boundary value problem

$$\begin{cases} \operatorname{div}(\mathbb{C}\widehat{\nabla}\boldsymbol{v}) = \boldsymbol{b} & \text{in } \mathbb{R}^3_- \\ \dfrac{\partial \boldsymbol{v}}{\partial \boldsymbol{\nu}} = \boldsymbol{0} & \text{on } \mathbb{R}^2 \\ \boldsymbol{v} = o(\boldsymbol{1}), \quad \nabla \boldsymbol{v} = o(|\boldsymbol{x}|^{-1}) & \text{as } |\boldsymbol{x}| \to +\infty. \end{cases} \tag{3.13}$$

The Neumann function of (3.13) is the kernel $\mathbf{N}$ of the integral operator

$$\boldsymbol{v}(\boldsymbol{x}) = \int_{\mathbb{R}^3_-} \mathbf{N}(\boldsymbol{x}, \boldsymbol{y}) \boldsymbol{b}(\boldsymbol{y})\, d\boldsymbol{y}, \tag{3.14}$$

which gives the solution to (3.13).

Given $\boldsymbol{y} = (y_1, y_2, y_3)$, we recall that $\widetilde{\boldsymbol{y}} := (y_1, y_2, -y_3)$.

Theorem 3.1.7 *The Neumann function* $\mathbf{N}$ *of problem* (3.13) *can be decomposed as*

$$\mathbf{N}(\boldsymbol{x}, \boldsymbol{y}) = \boldsymbol{\Gamma}(\boldsymbol{x} - \boldsymbol{y}) + \mathbf{R}^1(\boldsymbol{x} - \widetilde{\boldsymbol{y}}) + y_3 \mathbf{R}^2(\boldsymbol{x} - \widetilde{\boldsymbol{y}}) + y_3^2\, \mathbf{R}^3(\boldsymbol{x} - \widetilde{\boldsymbol{y}}),$$

where $\boldsymbol{\Gamma}$ *is the Kelvin matrix, see* (3.6), *and* $\mathbf{R}^k$, $k = 1, 2, 3$, *have components* R^k_{ij} *given by*

$$\begin{aligned} R^1_{ij}(\boldsymbol{\eta}) &:= C_{\mu,\nu}\big\{ -(\tilde{f} + c_\nu \tilde{g})\delta_{ij} - (3-4\nu)\eta_i\eta_j \tilde{f}^3 \\ &\qquad + c_\nu\big[\delta_{i3}\eta_j - \delta_{j3}(1-\delta_{i3})\eta_i\big]\tilde{f}\tilde{g} + c_\nu(1-\delta_{i3})(1-\delta_{j3})\eta_i\eta_j \tilde{f}\tilde{g}^2\big\} \\ R^2_{ij}(\boldsymbol{\eta}) &:= 2C_{\mu,\nu}\big\{(3-4\nu)\big[\delta_{i3}(1-\delta_{j3})\eta_j + \delta_{j3}(1-\delta_{i3})\eta_i\big]\tilde{f}^3 \\ &\qquad - (1-2\delta_{3j})\delta_{ij}\eta_3 \tilde{f}^3 + 3(1-2\delta_{3j})\eta_i\eta_j\eta_3 \tilde{f}^5\big\} \\ R^3_{ij}(\boldsymbol{\eta}) &:= 2C_{\mu,\nu}(1-2\delta_{j3})\big\{\delta_{ij}\tilde{f}^3 - 3\eta_i\eta_j \tilde{f}^5\big\}. \end{aligned}$$

for $i, j = 1, 2, 3$, *where* $c_\nu := 4(1-\nu)(1-2\nu)$, $C_{\mu,\nu} = \frac{1}{16\pi\mu(1-\nu)}$ *and*

$$\tilde{f}(\boldsymbol{\eta}) := \frac{1}{|\boldsymbol{\eta}|}, \qquad \tilde{g}(\boldsymbol{\eta}) := \frac{1}{|\boldsymbol{\eta}| - \eta_3}.$$

Moreover, for any $M_x, M_y > 0$, *such that* $M_x > M_y$, *there exists* $C > 0$ *such that*

$$|\mathbf{N}(\boldsymbol{x}, \boldsymbol{y})| \le C\,|\boldsymbol{x}|^{-1} \quad \text{and} \quad |\nabla \mathbf{N}(\boldsymbol{x}, \boldsymbol{y})| \le C\,|\boldsymbol{x}|^{-2} \tag{3.15}$$

for any $\boldsymbol{x}, \boldsymbol{y} \in \mathbb{R}^3_-$ *with* $|\boldsymbol{x}| \ge M_x$ *and* $|\boldsymbol{y}| \le M_y$.

We define

$$\mathbf{R}(\boldsymbol{\eta}, y_3) := \mathbf{R}^1(\boldsymbol{\eta}) + y_3\,\mathbf{R}^2(\boldsymbol{\eta}) + y_3^2\,\mathbf{R}^3(\boldsymbol{\eta}). \tag{3.16}$$

We notice that $\mathbf{R}$ is the regular part of the Neumann function since the singular point belongs to $\mathbb{R}^3_+$.

In order to prove this theorem, we recall the basic steps to deduce (3.14) using the potential approach in [113].

Papkovich–Neuber Potentials

The starting point is the Helmholtz decomposition of the vector field $\boldsymbol{v}$ in (3.13) as

$$\boldsymbol{v} = \nabla\phi + \nabla \times \boldsymbol{\psi}, \tag{3.17}$$

where ϕ is a scalar potential and $\boldsymbol{\psi}$ a vector potential. Since the divergence of $\boldsymbol{\psi}$ is arbitrary, $\boldsymbol{\psi}$ can be chosen in such a way that $\operatorname{div}\boldsymbol{\psi} = 0$. From the Lamé operator with volume forces $\boldsymbol{b}$ and the Helmholtz representation (3.17), we find that

$$\Delta\Big[\boldsymbol{v} + \frac{1}{(1-2\nu)}\nabla\phi\Big] = \frac{\boldsymbol{b}}{\mu}.$$

We define

$$\boldsymbol{h} := 4\pi\mu\Big[\boldsymbol{v} + \frac{1}{(1-2\nu)}\nabla\phi\Big], \tag{3.18}$$

where the constant $4\pi\mu$ has been added to simplify the calculations in the sequel, hence

$$\Delta\boldsymbol{h} = 4\pi\boldsymbol{b}, \qquad \operatorname{div}\boldsymbol{h} = \frac{8\pi\mu(1-\nu)}{1-2\nu}\Delta\phi. \tag{3.19}$$

By the identity $\Delta(\boldsymbol{x}\cdot\boldsymbol{h}) = \boldsymbol{x}\cdot\Delta\boldsymbol{h} + 2\operatorname{div}\boldsymbol{h}$ and the relation $\Delta\boldsymbol{h} = 4\pi\boldsymbol{b}$, we find that

$$\operatorname{div}\boldsymbol{h} = \tfrac{1}{2}\big[\Delta(\boldsymbol{x}\cdot\boldsymbol{h}) - 4\pi\boldsymbol{x}\cdot\boldsymbol{b}\big]. \tag{3.20}$$

Combining this expression with the second one in (3.19) we get

$$\Delta\left[\frac{8\pi\mu(1-\nu)}{1-2\nu}\phi - \frac{\boldsymbol{x}\cdot\boldsymbol{h}}{2}\right] = -2\pi\boldsymbol{x}\cdot\boldsymbol{b}.$$

We define the scalar quantity β as

$$\beta := \frac{16\pi\mu(1-\nu)}{1-2\nu}\phi - \boldsymbol{x}\cdot\boldsymbol{h}, \tag{3.21}$$

hence

$$\Delta\beta = -4\pi\boldsymbol{x}\cdot\boldsymbol{b}.$$

Using the definition (3.21) of β, we can avoid the dependence from ϕ into the relation (3.18), that is,

$$\boldsymbol{v} = C_{\mu,\nu}\{4(1-\nu)\boldsymbol{h} - \nabla(\beta + \boldsymbol{x}\cdot\boldsymbol{h})\}, \tag{3.22}$$

where $\boldsymbol{h}$ and β are the Papkovich–Neuber potentials.

Let us introduce the functions

$$\phi_0(\boldsymbol{x}) := \frac{1}{|\boldsymbol{x}|} \qquad \text{and} \qquad \psi_0(\boldsymbol{x}) := \frac{\phi_0(\boldsymbol{x})}{1 - x_3\phi_0(\boldsymbol{x})} = \frac{1}{|\boldsymbol{x}| - x_3},$$

observing that, apart from $\partial_i\phi_0 = -x_i\phi_0^3$, $i = 1, 2, 3$, the following identities hold, for $\alpha = 1, 2$,

$$\begin{aligned} \phi_0 - \psi_0 &= -x_3\phi_0\,\psi_0, & \partial_\alpha\psi_0 &= -x_\alpha\phi_0\,\psi_0^2,\\ \partial_3\psi_0 &= \phi_0\,\psi_0, & \partial_3(\phi_0\,\psi_0) &= \phi^3. \end{aligned}$$

We denote by ϕ and $\widetilde{\phi}$ the values $\phi_0(\boldsymbol{x} + \boldsymbol{e}_3)$ and $\phi_0(\boldsymbol{x} - \boldsymbol{e}_3)$, respectively, with analogous notation for ψ_0, i.e., ψ and $\widetilde{\psi}$ are the values $\psi_0(\boldsymbol{x} + \boldsymbol{e}_3)$ and $\psi_0(\boldsymbol{x} - \boldsymbol{e}_3)$, respectively.

Proposition 3.1.8 *Let* $\mathbf{I}$ *be the identity matrix and* δ *the Dirac delta concentrated at* $-\boldsymbol{e}_3$. *Then, the matrix-valued function* $\mathcal{N} = \mathcal{N}(\boldsymbol{x})$ *solution to*

$$\mathcal{L}\boldsymbol{v} := \operatorname{div}(\mathbb{C}\widehat{\nabla}\boldsymbol{v}) = \delta\mathbf{I} \quad \text{in } \mathbb{R}^3_-, \qquad (\mathbb{C}\widehat{\nabla}\boldsymbol{v})\boldsymbol{e}_3 = \mathbf{0} \quad \text{in } \mathbb{R}^2,$$

is given by

$$
\begin{aligned}
\mathcal{N}_{\alpha\alpha} &= -C_{\mu,\nu}\big\{(3-4\nu)\phi + x_\alpha^2\phi^3 + \widetilde{\phi} + [(3-4\nu)x_\alpha^2 - 2x_3]\widetilde{\phi}^3 + 6x_\alpha^2 x_3\widetilde{\phi}^5 \\
&\qquad + c_\nu\big(\widetilde{\psi} - x_\alpha^2\widetilde{\phi}\,\widetilde{\psi}^2\big)\big\} \\
\mathcal{N}_{\alpha\beta} &= -C_{\mu,\nu}x_\alpha x_\beta\big\{\phi^3 + (3-4\nu)\widetilde{\phi}^3 + 6x_3\widetilde{\phi}^5 - c_\nu\widetilde{\phi}\,\widetilde{\psi}^2\big\} \\
\mathcal{N}_{3\alpha} &= -C_{\mu,\nu}x_\alpha\big\{(x_3+1)\phi^3 + (3-4\nu)(x_3+1)\widetilde{\phi}^3 + 6x_3(x_3-1)\widetilde{\phi}^5 - c_\nu\widetilde{\phi}\,\widetilde{\psi}\big\} \\
\mathcal{N}_{\alpha 3} &= -C_{\mu,\nu}\,x_\alpha\big\{(x_3+1)\phi^3 + (3-4\nu)(x_3+1)\widetilde{\phi}^3 - 6x_3(x_3-1)\widetilde{\phi}^5 + c_\nu\widetilde{\phi}\,\widetilde{\psi}\big\} \\
\mathcal{N}_{33} &= -C_{\mu,\nu}\big\{(3-4\nu)\phi + (x_3+1)^2\phi^3 + (1+c_\nu)\widetilde{\phi} \\
&\qquad + \big[(3-4\nu)(x_3-1)^2 + 2x_3\big]\widetilde{\phi}^3 - 6x_3(x_3-1)^2\widetilde{\phi}^5\big\},
\end{aligned}
\tag{3.23}
$$

where $C_{\mu,\nu} = 1/\{16\pi\mu(1-\nu)\}$, $c_\nu = 4(1-\nu)(1-2\nu)$ *and* $\alpha = 1, 2$.

To prove the previous proposition and in particular (3.23), we observe that the columns $\mathcal{N}^{(i)}$ of $\mathcal{N}$ are determined by solving the equation $\mathcal{L}\boldsymbol{v} = \boldsymbol{e}_i\delta$ for $i = 1, 2, 3$. To solve this problem it is useful the Papkovich–Neuber representation (3.22) of $\boldsymbol{v}$, i.e.,

$$
\boldsymbol{v} = C_{\mu,\nu}\big\{4(1-\nu)\boldsymbol{h} - \nabla\big(\boldsymbol{x}\cdot\boldsymbol{h} + \beta\big)\big\} \quad \text{with} \quad \begin{cases} \Delta\boldsymbol{h} = 4\pi\,\boldsymbol{e}_i\delta \\ \Delta\beta = 4\pi\,\delta_{i3}\delta. \end{cases}
\tag{3.24}
$$

where δ_{ij} is the Kronecker symbol. The coupling between $\boldsymbol{h}$ and β is determined by the boundary conditions on the plane $\{x_3 = 0\}$, which are, for $\alpha = 1, 2$,

$$
\begin{aligned}
&(1-2\nu)(\partial_3 h_\alpha + \partial_\alpha h_3) - \boldsymbol{x}\cdot\partial^2_{\alpha 3}\boldsymbol{h} - \partial^2_{\alpha 3}\beta = 0, \\
&2\nu\,\mathrm{div}\,\boldsymbol{h} + 2(1-2\nu)\,\partial_3 h_3 - \boldsymbol{x}\cdot\partial^2_{33}\boldsymbol{h} - \partial^2_{33}\beta = 0.
\end{aligned}
\tag{3.25}
$$

Set

$$
G(\boldsymbol{x}, \boldsymbol{y}) := -\phi_0\big(\boldsymbol{x} - \boldsymbol{y}\big) + \phi_0\big(\boldsymbol{x} - \widetilde{\boldsymbol{y}}\big).
$$

We denote with $\langle f, g\rangle$ the duality pairing between f and the function g. Functions $\boldsymbol{h}$ and β can be determined by taking advantage of the relation (which descends from the second Green's identity)

$$
F(\boldsymbol{x}) = \tfrac{1}{4\pi}\langle \Delta F, G(\boldsymbol{x}, \cdot)\rangle,
\tag{3.26}
$$

applied to different choices of F, which will be specified step by step in the next proof.

Proof of Proposition 3.1.8 To determine $\mathcal{N}$, we consider separately the case of horizontal and vertical forcing. Due to the symmetry of the half-space, the role of x_1 and x_2 can be interchanged. Therefore we only prove the case of $\mathcal{L}\boldsymbol{v} = \boldsymbol{e}_1\delta$.

Horizontal Force $\mathcal{L}\boldsymbol{v} = \boldsymbol{e}_1\delta$. We choose $h_2 = 0$, so that boundary conditions become

$$\begin{cases} (1-2\nu)(\partial_3 h_1 + \partial_1 h_3) - x_1\partial_{13}^2 h_1 - \partial_{13}^2\beta = 0, \\ (1-2\nu)\partial_2 h_3 - x_1\partial_{23}^2 h_1 - \partial_{23}^2\beta = 0, & \text{for } x_3 = 0, \\ 2\nu\,\partial_1 h_1 + 2(1-\nu)\,\partial_3 h_3 - x_1\partial_{33}^2 h_1 - \partial_{33}^2\beta = 0. \end{cases}$$

Differentiating the first equation with respect to x_1, the second with respect to x_2 and taking the difference, we obtain

$$0 = (1-2\nu)\partial_{23}^2 h_1 + \partial_{23}^2 h_1 = 2(1-\nu)\partial_{23}^2 h_1 \qquad \text{for } x_3 = 0,$$

which suggests, after integration with respect to x_2, the choice $F := \partial_3 h_1$. Applying (3.26),

$$\partial_3 h_1 = -\partial_{y_3} G\big|_{y=-\boldsymbol{e}_3} = -\partial_3(\phi + \widetilde{\phi}), \qquad \text{for } x_3 < 0,$$

and thus $h_1 = -(\phi + \widetilde{\phi})$.

Being $\partial_3 h_1$ null for $x_3 = 0$, integration of the second boundary condition encourages the choice $F := (1-2\nu)h_3 - \partial_3\beta$ which is zero for $x_3 = 0$. Hence, since $\Delta F = 0$, we deduce

$$(1-2\nu)h_3 - \partial_3\beta = 0, \qquad \text{for } x_3 < 0. \tag{3.27}$$

Concerning the third boundary condition, for $x_3 = 0$ we observe that

$$\begin{aligned} \partial_1 h_1 &= x_1(\phi^3 + \widetilde{\phi}^3) = 2x_1\widetilde{\phi}^3 = -2\partial_1\widetilde{\phi} \\ x_1\partial_{33}^2 h_1 &= x_1(\phi^3 + \widetilde{\phi}^3 - 3\phi^5 - 3\widetilde{\phi}^5) = 2x_1(\widetilde{\phi}^3 - 3\widetilde{\phi}^5) = -2\big(\partial_1\widetilde{\phi} - \partial_{13}^2\widetilde{\phi}\big), \end{aligned}$$

since ϕ and $\widetilde{\phi}$ coincide when $x_3 = 0$. Substituting in the third boundary condition, we obtain

$$F := 2(1-\nu)\partial_3 h_3 - \partial_{33}^2\beta + 2(1-2\nu)\partial_1\widetilde{\phi} - 2\partial_{13}^2\widetilde{\phi} = 0 \quad \text{for } x_3 = 0.$$

Since $\Delta F = 0$, we infer

$$2(1-\nu)\partial_3 h_3 - \partial_{33}^2\beta + 2(1-2\nu)\partial_1\widetilde{\phi} - 2\partial_{13}^2\widetilde{\phi} = 0 \qquad \text{for } x_3 < 0,$$

and thus, being $\partial_1\widetilde{\phi} = -x_1\widetilde{\phi}^3 = -\partial_3(x_1\widetilde{\phi}\,\widetilde{\psi})$,

$$2(1-\nu)h_3 - \partial_3\beta = -2x_1\widetilde{\phi}^3 + 2(1-2\nu)x_1\widetilde{\phi}\,\widetilde{\psi} \qquad \text{for } x_3 < 0,$$

Coupling with (3.27), we deduce

$$\begin{cases} h_3 = -2x_1\widetilde{\phi}^3 + 2(1-2\nu)x_1\widetilde{\phi}\,\widetilde{\psi} \\ \partial_3\beta = -2(1-2\nu)x_1\widetilde{\phi}^3 + 2(1-2\nu)^2 x_1\widetilde{\phi}\,\widetilde{\psi} \end{cases} \qquad \text{for } x_3 < 0.$$

Recalling that $\widetilde{\phi}^3 = \partial_3(\widetilde{\phi}\,\widetilde{\psi})$ and $\widetilde{\phi}\,\widetilde{\psi} = \partial_3\widetilde{\psi}$, by integration,

$$\beta = -2(1-2\nu)x_1\widetilde{\phi}\,\widetilde{\psi} + 2(1-2\nu)^2 x_1\widetilde{\psi} \qquad \text{for } x_3 < 0.$$

Using the identity $(x_3-1)\widetilde{\phi}\,\widetilde{\psi} = \widetilde{\psi} - \widetilde{\phi}$, we infer

$$x_3h_3 + \beta = x_1\big\{-2(1-2\nu)\widetilde{\phi} - 2x_3\widetilde{\phi}^3 + c_\nu\widetilde{\psi}\big\}.$$

Substituting in (3.24), we get the expressions for $\mathcal{N}_{i1}$ given in (3.23).

Vertical Force $\mathcal{L}\boldsymbol{v} = \boldsymbol{e}_3\delta$. Choosing $h_1 = h_2 = 0$, conditions (3.25) become

$$\begin{cases} (1-2\nu)\partial_\alpha h_3 - \partial^2_{\alpha 3}\beta = 0 \qquad (\alpha = 1, 2), \\ 2(1-\nu)\partial_3 h_3 - \partial^2_{33}\beta = 0, \end{cases} \qquad \text{for } x_3 = 0.$$

Integrating the first relation with respect to x_α, we obtain

$$\begin{cases} (1-2\nu)h_3 - \partial_3\beta = 0, \\ 2(1-\nu)\partial_3 h_3 - \partial^2_{33}\beta = 0, \end{cases} \qquad \text{for } x_3 = 0.$$

Since $\Delta h_3 = \Delta\beta = \delta$, identity (3.26) with $F := (1-2\nu)h_3 - \partial_3\beta$ gives

$$\begin{aligned} (1-2\nu)h_3 - \partial_3\beta &= \big\{(1-2\nu)G + \partial_{y_3}G\big\}\big|_{y=-\boldsymbol{e}_3} \\ &= (1-2\nu)(-\phi + \widetilde{\phi}) - (x_3+1)\phi^3 - (x_3-1)\widetilde{\phi}^3, \end{aligned} \tag{3.28}$$

for $x_3 < 0$. Applying (3.26) to $F := 2(1-\nu)\partial_3 h_3 - \partial^2_{33}\beta$, we deduce

$$\begin{aligned} 2(1-\nu)\partial_3 h_3 - \partial^2_{33}\beta &= \big\{-2(1-\nu)\partial_{y_3}G - \partial^2_{y_3y_3}G\big\}\big|_{y=-\boldsymbol{e}_3} \\ &= \partial_3\big\{-2(1-\nu)(\phi + \widetilde{\phi}) + \partial_3(\phi - \widetilde{\phi})\big\}, \qquad \text{for } x_3 < 0. \end{aligned}$$

Integrating with respect to x_3, we infer

$$2(1-\nu)h_3 - \partial_3\beta = -2(1-\nu)(\phi + \widetilde{\phi}) - (x_3+1)\phi^3 + (x_3-1)\widetilde{\phi}^3, \qquad \text{for } x_3 < 0.$$

Coupling with (3.28), we get explicit expressions for h_3 and $\partial_3\beta$, namely

$$\begin{cases} h_3 = -\phi - (3-4\nu)\widetilde{\phi} + 2(x_3-1)\widetilde{\phi}^3, \\ \partial_3\beta = (x_3+1)\phi^3 - c_\nu\widetilde{\phi} + (3-4\nu)(x_3-1)\widetilde{\phi}^3 \end{cases} \quad \text{for } x_3 < 0.$$

Differentiation of $\partial_3\beta$ with respect to x_α gives

$$\begin{aligned} \partial^2_{3\alpha}\beta &= -3x_\alpha(x_3+1)\phi^5 + c_\nu x_\alpha\widetilde{\phi}^3 - 3(3-4\nu)x_\alpha(x_3-1)\widetilde{\phi}^5 \\ &= \partial_3\left\{x_\alpha\phi^3 + c_\nu x_\alpha\widetilde{\phi}\,\widetilde{\psi} + (3-4\nu)x_\alpha\widetilde{\phi}^3\right\}, \end{aligned}$$

and thus

$$\partial_\alpha\beta = x_\alpha\left\{\phi^3 + c_\nu\widetilde{\phi}\,\widetilde{\psi} + (3-4\nu)\widetilde{\phi}^3\right\}, \qquad \text{for } x_3 < 0.$$

Recalling identity (3.24), we deduce the corresponding expressions for $\mathcal{N}_{i3}$ in (3.23).

□

With the explicit expression of function $\mathcal{N}(\boldsymbol{x})$ at hand we can now prove Theorem 3.1.7.

Proof of Theorem 3.1.7 This proof is divided into three steps: Find the explicit expression of the Neumann function; prove its decay behavior at infinity; prove the uniqueness of (3.13).

Expression of the Neumann Function
The Neumann function $\mathbf{N} = \mathbf{N}(\boldsymbol{x}, \boldsymbol{y})$ in the half-space $\{x_3 < 0\}$ is such that its columns $\boldsymbol{N}_1$, $\boldsymbol{N}_2$, and $\boldsymbol{N}_3$ solve $\mathcal{L}\boldsymbol{N}_i = \delta_{\boldsymbol{y}}\boldsymbol{e}_i$, for $i = 1, 2, 3$, where δ_y is the Dirac delta concentrated at $\boldsymbol{y} = (y_1, y_2, y_3)$ with $y_3 < 0$. Thus, the Neumann function $\mathbf{N}$ is given by

$$\mathbf{N}(\boldsymbol{x}, \boldsymbol{y}) = \frac{1}{|y_3|}\mathcal{N}\left(\frac{x_1-y_1}{|y_3|}, \frac{x_2-y_2}{|y_3|}, \frac{x_3}{|y_3|}\right), \tag{3.29}$$

as a result of the homogeneity of δ and the second-order degree of $\mathcal{L}$.

Taking the definitions of $\phi, \widetilde{\phi}, \widetilde{\psi}$ and computing them at $(x_1 - y_1, x_2 - y_2, x_3)/|y_3|$, we obtain the identities

$$f := -\frac{\phi}{y_3} = \frac{1}{|\boldsymbol{x}-\boldsymbol{y}|}, \qquad \widetilde{f} := -\frac{\widetilde{\phi}}{y_3} = \frac{1}{|\boldsymbol{x}-\widetilde{\boldsymbol{y}}|},$$

$$\widetilde{g} := -\frac{\widetilde{\psi}}{y_3} = \frac{1}{|\boldsymbol{x}-\widetilde{\boldsymbol{y}}| - x_3 - y_3},$$

where $\widetilde{\mathbf{y}} = (y_1, y_2, -y_3)$. Hence, the components of $C^{-1}_{\mu,\nu}\mathbf{N}$ are given by

$$
\begin{aligned}
C^{-1}_{\mu,\nu}N_{\alpha\alpha} &= -(3-4\nu)f - (x_\alpha - y_\alpha)^2 f^3 - \tilde{f} - (3-4\nu)(x_\alpha - y_\alpha)^2\tilde{f}^3 - c_\nu\tilde{g} \\
&\quad + c_\nu(x_\alpha - y_\alpha)^2\tilde{f}\tilde{g}^2 - 2x_3y_3\tilde{f}^3 + 6(x_\alpha - y_\alpha)^2x_3y_3\tilde{f}^5 \\
C^{-1}_{\mu,\nu}N_{\alpha\beta} &= (x_\alpha - y_\alpha)(x_\beta - y_\beta)\big\{-f^3 - (3-4\nu)\tilde{f}^3 + c_\nu\tilde{f}\tilde{g}^2 + 6x_3y_3\tilde{f}^5\big\} \\
C^{-1}_{\mu,\nu}N_{3\alpha} &= (x_\alpha - y_\alpha)\big\{-(x_3 - y_3)f^3 - (3-4\nu)(x_3 - y_3)\tilde{f}^3 + c_\nu\tilde{f}\tilde{g} \\
&\quad + 6x_3y_3(x_3 + y_3)\tilde{f}^5\big\} \\
C^{-1}_{\mu,\nu}N_{\alpha 3} &= (x_\alpha - y_\alpha)\Big\{-(x_3 - y_3)f^3 - (3-4\nu)(x_3 - y_3)\tilde{f}^3 - c_\nu\tilde{f}\tilde{g} \\
&\quad - 6x_3y_3(x_3 + y_3)\tilde{f}^5\Big\} \\
C^{-1}_{\mu,\nu}N_{33} &= -(3-4\nu)f - (x_3 - y_3)^2 f^3 - (1 + c_\nu)\tilde{f} - (3-4\nu)(x_3 + y_3)^2\tilde{f}^3 \\
&\quad + 2x_3y_3\tilde{f}^3 - 6x_3y_3(x_3 + y_3)^2\tilde{f}^5.
\end{aligned}
$$

Using the expression of fundamental solution $\mathbf{\Gamma}$ for the Lamé operator, see (3.6), and the relation $\tilde{f} = \tilde{g} - (x_3 + y_3)\tilde{f}\tilde{g}$, the above formulas are rewritten as $\mathbf{N} = \mathbf{\Gamma} + \mathbf{R}$ where $\mathbf{\Gamma}$ is computed at $\boldsymbol{x} - \mathbf{y}$ and the component R_{ij}, for $i, j = 1, 2, 3$, of $\mathbf{R}$ are given by

$$
\begin{aligned}
R_{\alpha\alpha} &= C_{\mu,\nu}\big\{-(\tilde{f} + c_\nu\tilde{g}) - (3-4\nu)\eta_\alpha^2\tilde{f}^3 + c_\nu\eta_\alpha^2\tilde{f}\tilde{g}^2 - 2x_3y_3\big(\tilde{f}^3 - 3\eta_\alpha^2\tilde{f}^5\big)\big\} \\
R_{\beta\alpha} &= C_{\mu,\nu}\eta_\alpha\eta_\beta\big\{-(3-4\nu)\tilde{f}^3 + c_\nu\tilde{f}\tilde{g}^2 + 6x_3y_3\tilde{f}^5\big\} \\
R_{3\alpha} &= C_{\mu,\nu}\eta_\alpha\big\{-(3-4\nu)(\eta_3 - 2y_3)\tilde{f}^3 + c_\nu\tilde{f}\tilde{g} + 6x_3y_3\eta_3\tilde{f}^5\big\} \\
R_{\alpha 3} &= C_{\mu,\nu}\eta_\alpha\Big\{-(3-4\nu)(\eta_3 - 2y_3)\tilde{f}^3 - c_\nu\tilde{f}\tilde{g} - 6x_3y_3\eta_3\tilde{f}^5\Big\} \\
R_{33} &= C_{\mu,\nu}\big\{-(\tilde{f} + c_\nu\tilde{g}) - (3-4\nu)\eta_3^2\tilde{f}^3 + c_\nu\eta_3\tilde{f}\tilde{g} + 2x_3y_3\big(\tilde{f}^3 - 3\eta_3^2\tilde{f}^5\big)\big\},
\end{aligned}
$$

where $\eta_\alpha = x_\alpha - y_\alpha$ for $\alpha = 1, 2$ and $\eta_3 = x_3 + y_3$, which can be recombined as

$$
\begin{aligned}
R_{ij} = C_{\mu,\nu}\big\{&-(\tilde{f} + c_\nu\tilde{g})\delta_{ij} - (3-4\nu)\eta_i\eta_j\tilde{f}^3 \\
&+ 2(3-4\nu)y_3\big[\delta_{3i}(1-\delta_{3j})\eta_j + \delta_{3j}(1-\delta_{3i})\eta_i\big]\tilde{f}^3 \\
&+ c_\nu\big[\delta_{i3}\eta_j - \delta_{3j}(1-\delta_{3i})\eta_i\big]\tilde{f}\tilde{g} + c_\nu(1-\delta_{3j})(1-\delta_{3i})\eta_i\eta_j\tilde{f}\tilde{g}^2 \\
&- 2(1-2\delta_{3j})x_3y_3\big(\delta_{ij}\tilde{f}^3 - 3\eta_i\eta_j\tilde{f}^5\big)\big\},
\end{aligned}
$$

for $i, j = 1, 2, 3$. Since $x_3 = \eta_3 - y_3$, we obtain the decomposition $R_{ij} := R^1_{ij} + R^2_{ij} + R^3_{ij}$, where

$$R^1_{ij} := C_{\mu,\nu}\big\{ -(\tilde f + c_\nu \tilde g)\delta_{ij} - (3-4\nu)\eta_i\eta_j \tilde f^3 + c_\nu\big[\delta_{3i}\eta_j - \delta_{3j}(1-\delta_{3i})\eta_i\big]\tilde f \tilde g$$
$$+ c_\nu(1-\delta_{3j})(1-\delta_{3i})\eta_i\eta_j \tilde f \tilde g^2\big\}$$
$$R^2_{ij} := 2C_{\mu,\nu} y_3\big\{(3-4\nu)\big[\delta_{3i}(1-\delta_{3j})\eta_j + \delta_{3j}(1-\delta_{3i})\eta_i\big]\tilde f^3$$
$$-(1-2\delta_{3j})\delta_{ij}\eta_3 \tilde f^3 + 3(1-2\delta_{3j})\eta_i\eta_j\eta_3 \tilde f^5\big\}$$
$$R^3_{ij} := 2C_{\mu,\nu}(1-2\delta_{3j}) y_3^2\big\{\delta_{ij}\tilde f^3 - 3\eta_i\eta_j \tilde f^5\big\},$$

that is the assertion.

Behavior at Infinity
For the proof of the decay behavior at infinity we note that if ς is a homogeneous function of degree α which is defined and continuous in $\mathbb{R}^3_- \setminus \{\mathbf{0}\}$, then there exists a constant C such that

$$|\varsigma(\boldsymbol{x})| \le C|\boldsymbol{x}|^\alpha, \qquad \boldsymbol{x} \in \mathbb{R}^3_- \setminus \{\mathbf{0}\}.$$

Thus, since $\mathbf{R}^k$ are homogeneous of degree $-k$, for $k = 1, 2, 3$, and there exists $M_y > 0$ such that for any $\mathbf{y} \in \mathbb{R}^3_-$ with $|\mathbf{y}| \le M_y$

$$|\boldsymbol{\eta}| - \eta_3 \ge |\boldsymbol{\eta}| = |\boldsymbol{x} - \tilde{\boldsymbol{y}}| \ge |\boldsymbol{x}| - M_y,$$

for $|\boldsymbol{x}|$ sufficiently large, the term $\mathbf{R}$ is bounded by

$$|\mathbf{R}| \le |\mathbf{R}^1| + |y_3||\mathbf{R}^2| + |y_3|^2|\mathbf{R}^3| \le C\left(\frac{1}{|\boldsymbol{x}|} + \frac{|y_3|}{|\boldsymbol{x}|^2} + \frac{|y_3|^2}{|\boldsymbol{x}|^3}\right) \le \frac{C}{|\boldsymbol{x}|}.$$

Joining this bound with (3.8), we deduce the bound for $\mathbf{N}$.

The estimates on $|\nabla\mathbf{N}|$ is consequence of the homogeneity of derivatives of homogeneous functions together with the observation that $\tilde f$ and $\tilde g$ are C^1 in $\mathbb{R}^3_- \setminus \{\mathbf{0}\}$.

Uniqueness for (3.13)
The uniqueness of the solution of (3.13) can be proven by the energy method. Indeed, let $\boldsymbol{v}^1$ and $\boldsymbol{v}^2$ two solutions of (3.13). Their difference $\boldsymbol{v} := \boldsymbol{v}^1 - \boldsymbol{v}^2$ solves the homogeneous problem, i.e.,

$$\begin{cases} \operatorname{div}(\mathbb{C}\widehat{\nabla}\boldsymbol{v}) = \mathbf{0} & \text{in } \mathbb{R}^3_- \\ \dfrac{\partial \boldsymbol{v}}{\partial \boldsymbol{\nu}} = \mathbf{0} & \text{on } \mathbb{R}^2 \\ \boldsymbol{v} = O(|\boldsymbol{x}|^{-1}), \quad \nabla\boldsymbol{v} = o(|\boldsymbol{x}|^{-1}) & \text{as } |\boldsymbol{x}| \to +\infty, \end{cases} \tag{3.30}$$

where the behavior at infinity of $\boldsymbol{v}$ is given by the explicit decay condition of $\mathbf{N}$. Next we apply Betti's formula (3.4) to $\boldsymbol{v}$ in $B_r^-(\mathbf{0})$, with $r > 0$, i.e.,

$$\int_{\partial B_r^-(\mathbf{0})} \boldsymbol{v} \cdot \frac{\partial \boldsymbol{v}}{\partial \boldsymbol{\nu}} \, d\sigma(\boldsymbol{x}) = \int_{B_r^-(\mathbf{0})} \mathbb{C}\widehat{\nabla}\boldsymbol{v} : \widehat{\nabla}\boldsymbol{v} \, d\boldsymbol{x}.$$

On the boundary of the half-space $\partial \boldsymbol{v}/\partial \boldsymbol{\nu} = \mathbf{0}$; hence, utilizing the decay conditions at infinity for $\boldsymbol{v}$ and $\nabla \boldsymbol{v}$, we find that

$$\left| \int_{\partial B_r^b(\mathbf{0})} \boldsymbol{v} \cdot \frac{\partial \boldsymbol{v}}{\partial \boldsymbol{\nu}} \, d\sigma(\boldsymbol{x}) \right| \to 0, \qquad as\ r \to +\infty.$$

Therefore

$$\int_{\mathbb{R}^3_-} \mathbb{C}\widehat{\nabla}\boldsymbol{v} : \widehat{\nabla}\boldsymbol{v} \, d\boldsymbol{x} = 0,$$

so thanks to (3.3) we find that $\widehat{\nabla}\boldsymbol{v} = \mathbf{0}$ which implies that $\boldsymbol{v} = \boldsymbol{a} + \mathbf{A}\boldsymbol{x}$, where $\boldsymbol{a} \in \mathbb{R}^3$ and $\mathbf{A} \in \mathbb{R}^{3\times 3}$ is a skew matrix. On the other hand, by the decay conditions at infinity for $\boldsymbol{v}$ and $\nabla \boldsymbol{v}$ we find that $\boldsymbol{a} = \mathbf{0}$ and $\mathbf{A} = \mathbf{0}$. This concludes the proof since $\boldsymbol{v} = \mathbf{0}$. □

Now, we have all the instruments to analyze the elastic boundary value problem.

3.2 The Elastic Problem: Well-Posedness

In this section we analyze the boundary value problem presented at the beginning of this chapter, that is

$$\begin{cases} \operatorname{div}(\mathbb{C}\widehat{\nabla}\boldsymbol{u}) = \mathbf{0} & \text{in } \mathbb{R}^3_- \setminus C \\ \dfrac{\partial \boldsymbol{u}}{\partial \boldsymbol{\nu}} = p\,\boldsymbol{n} & \text{on } \partial C \\ \dfrac{\partial \boldsymbol{u}}{\partial \boldsymbol{\nu}} = \mathbf{0} & \text{on } \mathbb{R}^2 \\ \boldsymbol{u} = o(\mathbf{1}), \quad \nabla \boldsymbol{u} = o(|\boldsymbol{x}|^{-1}) & |\boldsymbol{x}| \to \infty, \end{cases} \tag{3.31}$$

where C is the cavity and p the pressure.

As did in the previous chapter, we show two ways of proving the well-posedness of this problem: one way is to study the well-posedness through integral equations, i.e., to provide an integral representation formula for solutions to (3.31) and then establishing the invertibility of some specific integral operators.

Another pathway is to employ the weighted Sobolev spaces and the weak formulation of (3.31). We stress that for the integral approach we adopt a different strategy with respect to the scalar case presented in Chap. 2. In fact, here, the integral representation formula needs to prove the well-posedness. To get the integral representation formula we need to use the expression of the Neumann function $\mathbf{N}$ provided in Theorem 3.1.7.

Both these methods have their own advantages: The integral representation formula is of help to find the asymptotic expansion of the solution of (3.31) (using the approach of Ammari and Kang, see [21]); the variational formulation of the problem is of help to solve the inverse problem of identifying the cavity from one measurement of the displacement provided on a portion of the boundary of the half-space. We will in detail discuss these two points next.

3.2.1 Well-Posedness via Integral Equations

In this section we derive an integral representation formula for $\boldsymbol{u}$ which satisfies problem (3.31). For this purpose, we utilize single and double layer potentials defined in (3.9), with the kernel given by the fundamental solution and, in addition, integral operators with the kernel given by the regular part $\mathbf{R}$ of the Neumann function, i.e.,

$$\begin{aligned} \mathbf{S}^R\boldsymbol{\varphi}(\boldsymbol{x}) &:= \int_{\partial C} (\mathbf{R}(\boldsymbol{x},\boldsymbol{y}))^T \boldsymbol{\varphi}(\boldsymbol{y})\, d\sigma(\boldsymbol{y}), && \boldsymbol{x} \in \mathbb{R}^3_-, \\ \mathbf{D}^R\boldsymbol{\varphi}(\boldsymbol{x}) &:= \int_{\partial C} \left(\frac{\partial \mathbf{R}}{\partial \boldsymbol{\nu}(\boldsymbol{y})}(\boldsymbol{x},\boldsymbol{y})\right)^T \boldsymbol{\varphi}(\boldsymbol{y})\, d\sigma(\boldsymbol{y}), && \boldsymbol{x} \in \mathbb{R}^3_-, \end{aligned} \tag{3.32}$$

where $\boldsymbol{\varphi} \in \boldsymbol{L}^2(\partial C)$.

Theorem 3.2.1 *The solution $\boldsymbol{u}$ to* (3.31) *is such that*

$$\boldsymbol{u} = p\mathbf{S}^\Gamma \boldsymbol{n} - \mathbf{D}^\Gamma \boldsymbol{f} + p\mathbf{S}^R \boldsymbol{n} - \mathbf{D}^R \boldsymbol{f}, \qquad \text{in } \mathbb{R}^3_- \setminus \overline{C} \tag{3.33}$$

where $\mathbf{S}^\Gamma$, $\mathbf{D}^\Gamma$ are defined in (3.9), *$\mathbf{S}^R$, $\mathbf{D}^R$ in* (3.32), *$p\boldsymbol{n}$ is the boundary condition in* (3.31) *and $\boldsymbol{f}$ is the trace of $\boldsymbol{u}$ on ∂C, solution to*

$$\left(\tfrac{1}{2}\mathbf{I} + \mathbf{K} + \mathbf{D}^R\right)\boldsymbol{f} = p\left(\mathbf{S}^\Gamma \boldsymbol{n} + \mathbf{S}^R \boldsymbol{n}\right), \qquad \text{on } \partial C. \tag{3.34}$$

Proof of Theorem 3.2.1 Given $r, \varepsilon > 0$ such that $C \subset B_r^-(\mathbf{0})$ and $B_\varepsilon(\boldsymbol{y}) \subset \mathbb{R}^3_- \setminus \overline{C}$, let

$$\Omega_{r,\varepsilon} = \left(\mathbb{R}^3_- \cap B_r^-(\mathbf{0})\right) \setminus (C \cup B_\varepsilon(\boldsymbol{y})),$$

with r sufficiently large such that to contain the cavity C; additionally, we recall that $\partial B_r^h(\mathbf{0})$ is the intersection of the half ball $B_r^-(\mathbf{0})$ with the boundary of the half-space, and $\partial B_r^b(\mathbf{0})$ is the spherical cap (see Fig. 2.2, where now $d = 3$). In this domain, we apply Betti's formula (3.5) to $\boldsymbol{u}$ and the k-th column vector of $\mathbf{N}$, indicated by $\boldsymbol{N}^{(k)}$, for $k = 1, 2, 3$, in $\Omega_{r,\varepsilon}$, hence

$$
\begin{aligned}
0 &= \int_{\Omega_{r,\varepsilon}} \left[\boldsymbol{u}(\boldsymbol{x}) \cdot \mathcal{L}\boldsymbol{N}^{(k)}(\boldsymbol{x}, \boldsymbol{y}) - \boldsymbol{N}^{(k)}(\boldsymbol{x}, \boldsymbol{y}) \cdot \mathcal{L}\boldsymbol{u}(\boldsymbol{x}) \right] d\boldsymbol{x} \\
&= \int_{\partial B_r^b(\boldsymbol{0})} \left[\frac{\partial \boldsymbol{N}^{(k)}}{\partial \boldsymbol{\nu}_x}(\boldsymbol{x}, \boldsymbol{y}) \cdot \boldsymbol{u}(\boldsymbol{x}) - \boldsymbol{N}^{(k)}(\boldsymbol{x}, \boldsymbol{y}) \cdot \frac{\partial \boldsymbol{u}}{\partial \boldsymbol{\nu}_x}(\boldsymbol{x}) \right] d\sigma(\boldsymbol{x}) \\
&\quad - \int_{\partial B_\varepsilon(\boldsymbol{y})} \left[\frac{\partial \boldsymbol{N}^{(k)}}{\partial \boldsymbol{\nu}_x}(\boldsymbol{x}, \boldsymbol{y}) \cdot \boldsymbol{u}(\boldsymbol{x}) - \boldsymbol{N}^{(k)}(\boldsymbol{x}, \boldsymbol{y}) \cdot \frac{\partial \boldsymbol{u}}{\partial \boldsymbol{\nu}_x}(\boldsymbol{x}) \right] d\sigma(\boldsymbol{x}) \\
&\quad - \int_{\partial C} \left[\frac{\partial \boldsymbol{N}^{(k)}}{\partial \boldsymbol{\nu}_x}(\boldsymbol{x}, \boldsymbol{y}) \cdot \boldsymbol{u}(\boldsymbol{x}) - \boldsymbol{N}^{(k)}(\boldsymbol{x}, \boldsymbol{y}) \cdot \frac{\partial \boldsymbol{u}}{\partial \boldsymbol{\nu}_x}(\boldsymbol{x}) \right] d\sigma(\boldsymbol{x}) \\
&:= I_1 + I_2 + I_3,
\end{aligned}
$$

since, from (3.31) and the boundary condition in (3.13),

$$
\int_{\partial B_r^h(\boldsymbol{0})} \left[\frac{\partial \boldsymbol{N}^{(k)}}{\partial \boldsymbol{\nu}_x}(\boldsymbol{x}, \boldsymbol{y}) \cdot \boldsymbol{u}(\boldsymbol{x}) - \boldsymbol{N}^{(k)}(\boldsymbol{x}, \boldsymbol{y}) \cdot \frac{\partial \boldsymbol{u}}{\partial \boldsymbol{\nu}_x}(\boldsymbol{x}) \right] d\sigma(\boldsymbol{x}) = 0.
$$

The first step is to show that the term I_1 goes to zero. This follows by the behavior at infinity of $\boldsymbol{u}$ given in (3.31) and of the Neumann function given in (3.15). Indeed, we have

$$
\begin{aligned}
\left| \int_{\partial B_r^b(\boldsymbol{0})} \frac{\partial \boldsymbol{N}^{(k)}}{\partial \boldsymbol{\nu}_x}(\boldsymbol{x}, \boldsymbol{y}) \cdot \boldsymbol{u}(\boldsymbol{x}) \, d\sigma(\boldsymbol{x}) \right| &\leq \int_{\partial B_r^b(\boldsymbol{0})} |\boldsymbol{u}| \left| \frac{\partial \boldsymbol{N}^{(k)}}{\partial \boldsymbol{\nu}_x} \right| d\sigma(\boldsymbol{x}) \\
&\leq \frac{C}{r^2} \int_{\partial B_r^b(\boldsymbol{0})} |\boldsymbol{u}(\boldsymbol{x})| d\sigma(\boldsymbol{x}).
\end{aligned}
$$

This last integral can be estimated by means of the spherical coordinates $x_1 = r \sin\varphi \cos\theta$, $x_2 = r \sin\varphi \sin\theta$, $x_3 = r \cos\varphi$ where $\varphi \in [\pi/2, \pi]$, since $\partial B_r^b(\boldsymbol{0})$ is a spherical cap in $\mathbb{R}^3_-$, and $\theta \in [0, 2\pi)$, indeed

$$
\begin{aligned}
\frac{C}{r^2} \int_{\partial B_r^b(\boldsymbol{0})} |\boldsymbol{u}| \, d\sigma(\boldsymbol{x}) &= C \int_{\frac{\pi}{2}}^{\pi} \int_0^{2\pi} |\boldsymbol{u}(r, \theta, \varphi)| \sin\varphi \, d\theta \, d\varphi \\
&\leq C \sup_{\theta \in [0, 2\pi), \varphi \in [\frac{\pi}{2}, \pi]} |\boldsymbol{u}(r, \theta, \varphi)| \to 0,
\end{aligned}
$$

as $r \to +\infty$, since $\boldsymbol{u} = o(\mathbf{1})$. Similarly

$$\left| \int_{\partial B_r^b(\mathbf{0})} \boldsymbol{N}^{(k)}(\boldsymbol{x}, \boldsymbol{y}) \cdot \frac{\partial \boldsymbol{u}}{\partial \boldsymbol{\nu}_{\boldsymbol{x}}}(\boldsymbol{x})\, d\sigma(\boldsymbol{x}) \right| \leq \int_{\partial B_r^b(\mathbf{0})} |\boldsymbol{N}^{(k)}| \left| \frac{\partial \boldsymbol{u}}{\partial \boldsymbol{\nu}_{\boldsymbol{x}}} \right| d\sigma(\boldsymbol{x})$$
$$\leq \frac{C}{r} \int_{\partial B_r^b} \left| \frac{\partial \boldsymbol{u}}{\partial \boldsymbol{\nu}_{\boldsymbol{x}}} \right| d\sigma(\boldsymbol{x}).$$

By utilizing spherical coordinates again, we get

$$\frac{C}{r} \int_{\partial B_r^b(\mathbf{0})} \left| \frac{\partial \boldsymbol{u}}{\partial \boldsymbol{\nu}_{\boldsymbol{x}}} \right| d\sigma(\boldsymbol{x}) \leq C \sup_{\theta \in [0, 2\pi), \varphi \in [\frac{\pi}{2}, \pi]} r \left| \frac{\partial \boldsymbol{u}}{\partial \boldsymbol{\nu}}(r, \theta, \varphi) \right| \to 0, \tag{3.35}$$

as $r \to +\infty$, since $|\nabla \boldsymbol{u}| = o(r^{-1})$.

Integral I_2 gives the value of the function $\boldsymbol{u}$ in $\boldsymbol{y}$ as ε goes to zero. Indeed, we have

$$\left| \int_{\partial B_\varepsilon(\boldsymbol{y})} \boldsymbol{N}^{(k)}(\boldsymbol{x}, \boldsymbol{y}) \cdot \frac{\partial \boldsymbol{u}}{\partial \boldsymbol{\nu}_{\boldsymbol{x}}}(\boldsymbol{x})\, d\sigma(\boldsymbol{x}) \right| \leq \int_{\partial B_\varepsilon(\boldsymbol{y})} |\boldsymbol{N}^{(k)}| \left| \frac{\partial \boldsymbol{u}}{\partial \boldsymbol{\nu}_{\boldsymbol{x}}} \right| d\sigma(\boldsymbol{x})$$
$$\leq \sup_{\boldsymbol{x} \in \partial B_\varepsilon(\boldsymbol{y})} \left| \frac{\partial \boldsymbol{u}}{\partial \boldsymbol{\nu}_{\boldsymbol{x}}} \right| \int_{\partial B_\varepsilon(\boldsymbol{y})} \left[|\boldsymbol{\Gamma}^{(k)}| + |\boldsymbol{R}^{(k)}| \right] d\sigma(\boldsymbol{x}) = O(\varepsilon),$$

since the second integral has a continuous kernel. On the other hand

$$-\int_{\partial B_\varepsilon(\boldsymbol{y})} \frac{\partial \boldsymbol{N}^{(k)}}{\partial \boldsymbol{\nu}_{\boldsymbol{x}}}(\boldsymbol{x}, \boldsymbol{y}) \cdot \boldsymbol{u}(\boldsymbol{x})\, d\sigma(\boldsymbol{x}) = -\boldsymbol{u}(\boldsymbol{y}) \cdot \int_{\partial B_\varepsilon(\boldsymbol{y})} \frac{\partial \boldsymbol{N}^{(k)}}{\partial \boldsymbol{\nu}_{\boldsymbol{x}}}(\boldsymbol{x}, \boldsymbol{y})\, d\sigma(\boldsymbol{x})$$
$$+ \int_{\partial B_\varepsilon(\boldsymbol{y})} [\boldsymbol{u}(\boldsymbol{y}) - \boldsymbol{u}(\boldsymbol{x})] \cdot \frac{\partial \boldsymbol{N}^{(k)}}{\partial \boldsymbol{\nu}_{\boldsymbol{x}}}(\boldsymbol{x}, \boldsymbol{y})\, d\sigma(\boldsymbol{x}).$$

The latter integral tends to zero when ε goes to zero because

$$\left| \int_{\partial B_\varepsilon(\boldsymbol{y})} [\boldsymbol{u}(\boldsymbol{y}) - \boldsymbol{u}(\boldsymbol{x})] \cdot \frac{\partial \boldsymbol{N}^{(k)}}{\partial \boldsymbol{\nu}_{\boldsymbol{x}}}(\boldsymbol{x}, \boldsymbol{y})\, d\sigma(\boldsymbol{x}) \right|$$
$$\leq \sup_{\boldsymbol{x} \in \partial B_\varepsilon(\boldsymbol{y})} |\boldsymbol{u}(\boldsymbol{y}) - \boldsymbol{u}(\boldsymbol{x})| \int_{\partial B_\varepsilon(\boldsymbol{y})} \left| \frac{\partial \boldsymbol{N}^{(k)}}{\partial \boldsymbol{\nu}_{\boldsymbol{x}}} \right| d\sigma(\boldsymbol{x}),$$

and this last integral is bounded when ε goes to zero. We finally observe that

$$-\boldsymbol{u}(\boldsymbol{y}) \cdot \int_{\partial B_\varepsilon(\boldsymbol{y})} \frac{\partial \boldsymbol{N}^{(k)}}{\partial \boldsymbol{\nu}_{\boldsymbol{x}}}(\boldsymbol{x}, \boldsymbol{y})\, d\sigma(\boldsymbol{x}) = -\boldsymbol{u}(\boldsymbol{y}) \cdot \int_{\partial B_\varepsilon(\boldsymbol{y})} \frac{\partial (\boldsymbol{\Gamma}^{(k)} + \boldsymbol{R}^{(k)})}{\partial \boldsymbol{\nu}_{\boldsymbol{x}}}(\boldsymbol{x}, \boldsymbol{y})\, d\sigma(\boldsymbol{x})$$

$$= -\boldsymbol{u}(\boldsymbol{y}) \cdot \int_{\partial B_\varepsilon(\boldsymbol{y})} \frac{\partial \boldsymbol{\Gamma}^{(k)}}{\partial \boldsymbol{\nu}_{\boldsymbol{x}}}(\boldsymbol{x} - \boldsymbol{y})\, d\sigma(\boldsymbol{x}) - \boldsymbol{u}(\boldsymbol{y}) \cdot \int_{\partial B_\varepsilon(\boldsymbol{y})} \frac{\partial \boldsymbol{R}^{(k)}}{\partial \boldsymbol{\nu}_{\boldsymbol{x}}}(\boldsymbol{x}, \boldsymbol{y})\, d\sigma(\boldsymbol{x}), \tag{3.36}$$

where the latter integral tends to zero as $\varepsilon \to 0$, since $\boldsymbol{R}^{(k)}$, for $k = 1, 2, 3$, is the regular part of the Neumann function. For the first integral, we preliminarily observe that direct differentiation gives

$$\left(\frac{\partial \boldsymbol{\Gamma}^{(k)}}{\partial \boldsymbol{\nu}_{\boldsymbol{x}}}\right)_h (\boldsymbol{x} - \boldsymbol{y}) = -c'_\nu \left\{ n_k(\boldsymbol{x}) \frac{\partial}{\partial x_h} \frac{1}{|\boldsymbol{x} - \boldsymbol{y}|} - n_h(\boldsymbol{x}) \frac{\partial}{\partial x_k} \frac{1}{|\boldsymbol{x} - \boldsymbol{y}|} \right.$$
$$\left. + \left[\delta_{hk} + \frac{3}{(1 - 2\nu)} \frac{\partial |\boldsymbol{x} - \boldsymbol{y}|}{\partial x_k} \frac{\partial |\boldsymbol{x} - \boldsymbol{y}|}{\partial x_h} \right] \frac{\partial}{\partial \boldsymbol{n}(\boldsymbol{x})} \frac{1}{|\boldsymbol{x} - \boldsymbol{y}|} \right\}, \tag{3.37}$$

where $c'_\nu := (1 - 2\nu)/(8\pi(1 - \nu))$.

We substitute this expression into the integral (3.36) and we take into account that

$$n_h(\boldsymbol{x}) = \frac{x_h - y_h}{|\boldsymbol{x} - \boldsymbol{y}|}, \qquad \frac{\partial}{\partial x_k} \frac{1}{|\boldsymbol{x} - \boldsymbol{y}|} = -\frac{x_k - y_k}{|\boldsymbol{x} - \boldsymbol{y}|^3},$$

hence

$$\int_{\partial B_\varepsilon(\boldsymbol{y})} n_h(\boldsymbol{x}) \frac{\partial}{\partial x_k} \frac{1}{|\boldsymbol{x} - \boldsymbol{y}|}\, d\sigma(\boldsymbol{x}) = -\int_{\partial B_\varepsilon(\boldsymbol{y})} \frac{(x_h - y_h)(x_k - y_k)}{|\boldsymbol{x} - \boldsymbol{y}|^4}\, d\sigma(\boldsymbol{x}).$$

To solve this last integral we use spherical coordinates, that is,

$$x_1 - y_1 = \varepsilon \sin\varphi \cos\theta, \qquad x_2 - y_2 = \varepsilon \sin\varphi \sin\theta, \qquad x_3 - y_3 = \varepsilon \cos\varphi,$$

where $\varphi \in [0, \pi]$ and $\theta \in [0, 2\pi)$. From a straightforward calculation it follows

$$-\int_{\partial B_\varepsilon(\boldsymbol{y})} \frac{(x_h - y_h)(x_k - y_k)}{|\boldsymbol{x} - \boldsymbol{y}|^4}\, d\sigma(\boldsymbol{x}) = \begin{cases} 0 & \text{if } h \neq k \\ -\frac{4}{3}\pi & \text{if } h = k. \end{cases} \tag{3.38}$$

Therefore, from (3.37) and (3.38), we have

$$\int_{\partial B_\varepsilon(\boldsymbol{y})} \left(n_k(\boldsymbol{x}) \frac{\partial}{\partial x_h} \frac{1}{|\boldsymbol{x}-\boldsymbol{y}|} - n_h(\boldsymbol{x}) \frac{\partial}{\partial x_k} \frac{1}{|\boldsymbol{x}-\boldsymbol{y}|} \right) d\sigma(\boldsymbol{x}) = 0, \tag{3.39}$$

for any h and k. Hence, (3.36) becomes

$$\begin{aligned} &- \boldsymbol{u}(\boldsymbol{y}) \cdot \int_{\partial B_\varepsilon(\boldsymbol{y})} \frac{\partial \boldsymbol{N}^{(k)}}{\partial \boldsymbol{\nu}_{\boldsymbol{x}}}(\boldsymbol{x}-\boldsymbol{y})\, d\sigma(\boldsymbol{x}) \\ &= c'_\nu \sum_{h=1}^{3} u_h(\boldsymbol{y}) \int_{\partial B_\varepsilon(\boldsymbol{y})} \left(\delta_{hk} + \frac{3}{(1-2\nu)} \frac{\partial|\boldsymbol{x}-\boldsymbol{y}|}{\partial x_k} \frac{\partial|\boldsymbol{x}-\boldsymbol{y}|}{\partial x_h} \right) \frac{\partial}{\partial \boldsymbol{n}_{\boldsymbol{x}}} \frac{1}{|\boldsymbol{x}-\boldsymbol{y}|}\, d\sigma(\boldsymbol{x}) \\ &\qquad + O(\varepsilon). \end{aligned}$$

Utilizing the spherical coordinates another time and the definition of c'_ν, we find that

$$\frac{1-2\nu}{8\pi(1-\nu)} \int_{\partial B_\varepsilon(\boldsymbol{y})} \delta_{hk} \frac{\partial}{\partial \boldsymbol{n}_{\boldsymbol{x}}} \frac{1}{|\boldsymbol{x}-\boldsymbol{y}|}\, d\sigma(\boldsymbol{x}) = \begin{cases} -\frac{1-2\nu}{2(1-\nu)} & \text{if } h = k \\ 0 & \text{if } h \neq k. \end{cases} \tag{3.40}$$

Similarly

$$\begin{aligned} &\frac{3}{8\pi(1-\nu)} \int_{\partial B_\varepsilon(\boldsymbol{y})} \left(\frac{\partial|\boldsymbol{x}-\boldsymbol{y}|}{\partial x_k} \frac{\partial|\boldsymbol{x}-\boldsymbol{y}|}{\partial x_h} \right) \frac{\partial}{\partial \boldsymbol{n}_{\boldsymbol{x}}} \frac{1}{|\boldsymbol{x}-\boldsymbol{y}|}\, d\sigma(\boldsymbol{x}) \\ &\qquad = \begin{cases} -\frac{1}{2(1-\nu)} & \text{if } h = k \\ 0 & \text{if } h \neq k. \end{cases} \end{aligned} \tag{3.41}$$

Lastly from (3.40) and (3.41), we find that

$$\lim_{\varepsilon \to 0} \left(-\boldsymbol{u}(\boldsymbol{y}) \cdot \int_{\partial B_\varepsilon(\boldsymbol{y})} \frac{\partial \boldsymbol{N}^{(k)}}{\partial \boldsymbol{\nu}_{\boldsymbol{x}}}(\boldsymbol{x}-\boldsymbol{y})\, d\sigma(\boldsymbol{x}) \right) = -u_k(\boldsymbol{y}).$$

Using the definition of single and double layer potentials (3.9), (3.32) and splitting $\mathbf{N}$ as $\boldsymbol{\Gamma} + \mathbf{R}$ formula (3.33) holds.

To find the integral equation satisfied by f is sufficient to apply the trace properties of the double layer potentials (3.10) in formula (3.33). □

By means of the integral representation formula (3.33) and the behaviour of the Neumann function at infinity, see (3.15), it immediately follows

Corollary 3.2.2 *If $\boldsymbol{u}$ is a solution to* (3.31)*, then*

$$\boldsymbol{u}(\boldsymbol{y}) = O(|\boldsymbol{y}|^{-1}) \qquad as \quad |\boldsymbol{y}| \to \infty. \tag{3.42}$$

The well-posedness of the boundary value problem (3.31) is reduced to prove the invertibility of

$$\tfrac{1}{2}\mathbf{I} + \mathbf{K} + \mathbf{D}^R : \boldsymbol{L}^2(\partial C) \to \boldsymbol{L}^2(\partial C), \tag{3.43}$$

which corresponds to show the injectivity and surjectivity of the above operator. Specifically, the injectivity follows by proving the uniqueness of $\boldsymbol{u}$ which can be obtained by the classical energy method, see [83, 104]. Once we have the injectivity, the proof of the surjectivity of (3.43) follows by the application of the index theorem regarding bounded and linear operators.

For this purpose, we here recall three important theorems of the functional analysis for reader's convenience: the Ascoli-Arzelà theorem, the closed range theorem and the bounded inverse theorem (see [53, 99, 103, 138]). They are useful to collect some information on the integral operators $1/2\mathbf{I} + \mathbf{K}$ and $\mathbf{D}^R$.

Theorem 3.2.3 ([103]) *Let $\Omega \subset \mathbb{R}^d$ be a compact set and $C(\Omega)$ the space of continuous functions defined on Ω. A set $S \subset C(\Omega)$ is relatively compact if and only if it is bounded and equicontinuous, i.e., if there exists for all $\boldsymbol{x} \in \Omega$ and all $\psi \in S$ a constant c such that*

$$|\psi(\boldsymbol{x})| \leq c, \tag{3.44}$$

and for all $\varepsilon > 0$ there exists $\delta > 0$ such that

$$|\psi(\boldsymbol{x}) - \psi(\boldsymbol{y})| < \varepsilon$$

for all $\boldsymbol{x}, \boldsymbol{y} \in \Omega$ such that $|\boldsymbol{x} - \boldsymbol{y}| < \delta$, and all $\psi \in S$.

Theorem 3.2.4 ([99, 138]) *Let X and Y be Banach spaces, and T a closed linear operator defined in X into Y such that $\overline{D(T)} = X$. Then the following propositions are all equivalent:*

a. $Im(T)$ is closed in Y; *b. $Im(T^*)$ is closed in X^*;*
c. $Im(T) = (Ker(T^*))^{\perp}$; *d. $Im(T^*) = (Ker(T))^{\perp}$.*

The next theorem, i.e., the bounded inverse theorem is a consequence of the open mapping theorem.

Theorem 3.2.5 ([53]) *Let X and Y be two Banach spaces and T a continuous linear operator from X into Y which is bijective. Then T^{-1} is also bounded.*

From the previous theorems, the next result follows

Lemma 3.2.6 *The operator $\frac{1}{2}\mathbf{I} + \mathbf{K} : \boldsymbol{L}^2(\partial C) \to \boldsymbol{L}^2(\partial C)$ is invertible with bounded inverse.*

Proof In [66] the authors proved that

$$\tfrac{1}{2}\mathbf{I} + \mathbf{K}^* : \boldsymbol{L}^2(\partial C) \to \boldsymbol{L}^2(\partial C)$$

is a bounded linear operator, injective, and with dense and closed range, i.e., is invertible. Therefore, from Theorem 3.2.4 we have

$$\mathrm{Ker}\left(\tfrac{1}{2}\mathbf{I} + \mathbf{K}\right) = \{0\}, \qquad \mathrm{Im}\left(\tfrac{1}{2}\mathbf{I} + \mathbf{K}\right)^{\perp} = \{0\}$$

and $\mathrm{Im}(1/2\mathbf{I} + \mathbf{K})$ is closed. From this considerations, we have that the operator $\frac{1}{2}\mathbf{I} + \mathbf{K} : \boldsymbol{L}^2(\partial C) \to \boldsymbol{L}^2(\partial C)$ is bijective. The assertion of the theorem follows utilizing Theorem 3.2.5. □

It is straightforward to show that the operator $\mathbf{D}^R$ is compact since it has a continuous kernel. For this purpose we adapt the arguments contained in [103].

Lemma 3.2.7 *The operator* $\mathbf{D}^R : \boldsymbol{L}^2(\partial C) \to \boldsymbol{L}^2(\partial C)$ *is compact.*

Proof For simplicity, we denote

$$\mathbf{H}(\boldsymbol{x}, \boldsymbol{y}) := \frac{\partial \mathbf{R}}{\partial \boldsymbol{\nu}}(\boldsymbol{x}, \boldsymbol{y}), \qquad \boldsymbol{x}, \boldsymbol{y} \in \partial C.$$

To show the compactness property we work columnwise with $\mathbf{H}$, i.e., we consider $\boldsymbol{H}^{(k)}$, for $k = 1, 2, 3$.

Let $\boldsymbol{S}$ be a bounded set such that $\boldsymbol{S} \subset \boldsymbol{L}^2(\partial C)$, that is $\|\boldsymbol{\varphi}\|_{\boldsymbol{L}^2(\partial C)} \leq K$, for any $\boldsymbol{\varphi} \in \boldsymbol{S}$. From the Cauchy–Schwarz inequality we get

$$|(\boldsymbol{D}^R \boldsymbol{\varphi}(\boldsymbol{y}))_k|^2 \leq \|\boldsymbol{H}^{(k)}(\cdot, \boldsymbol{y})\|^2_{\boldsymbol{L}^2(\partial C)} \|\boldsymbol{\varphi}\|^2_{\boldsymbol{L}^2(\partial C)} \leq K|\partial C| \max_{\boldsymbol{x}, \boldsymbol{y} \in \partial C} |\boldsymbol{H}^{(k)}|,$$

with $k = 1, 2, 3$, for all $\boldsymbol{y} \in \partial C$ and $\boldsymbol{\varphi} \in \boldsymbol{S}$. Hence $\|\boldsymbol{D}^R(\boldsymbol{\varphi})\| \leq K'$, with $K' > 0$, which implies that $\boldsymbol{D}^R(\boldsymbol{S})$ is bounded. Moreover, for all $\varepsilon > 0$ there exist $\boldsymbol{\varphi}, \boldsymbol{\varphi}' \in \boldsymbol{S}$ and $\delta > 0$ such that if $\|\boldsymbol{\varphi}(\boldsymbol{y}) - \boldsymbol{\varphi}'(\boldsymbol{y})\|_{\boldsymbol{L}^2(\partial C)} < \delta$ then, applying again the Cauchy–Schwarz inequality

$$|\boldsymbol{D}^R(\boldsymbol{\varphi} - \boldsymbol{\varphi}')(\boldsymbol{y})| < \varepsilon.$$

From this last result it follows that $\mathbf{D}^R(\boldsymbol{S})$ is an element of the space of continuous function on ∂C. Moreover, each component of the matrix $\mathbf{H}$ is uniformly continuous on the compact set $\partial C \times \partial C$, so for every $\varepsilon > 0$ there exists $\delta > 0$ such that

$$\|\boldsymbol{H}^{(k)}(z, \boldsymbol{x}) - \boldsymbol{H}^{(k)}(z, \boldsymbol{y})\| \leq \frac{\varepsilon}{\sqrt{3}K|\partial C|^{1/2}},$$

for all $\boldsymbol{x}, \boldsymbol{y}, z \in \partial C$ with $\|\boldsymbol{x} - \boldsymbol{y}\| < \delta$. Since

$$|(\boldsymbol{D}^R\boldsymbol{\varphi})_k(\boldsymbol{x}) - (\boldsymbol{D}^R\boldsymbol{\varphi})_k(\boldsymbol{y})| \leq \int_{\partial C} \|\boldsymbol{H}^{(k)}(z,\boldsymbol{x}) - \boldsymbol{H}^{(k)}(z,\boldsymbol{y})\| \|\boldsymbol{\varphi}(z)\| \, d\sigma(z)$$
$$\leq \frac{\varepsilon}{\sqrt{3}},$$

for $k = 1, 2, 3$, hence

$$\|(\boldsymbol{D}^R\boldsymbol{\varphi})(\boldsymbol{x}) - (\boldsymbol{D}^R\boldsymbol{\varphi})(\boldsymbol{y})\| \leq \varepsilon,$$

for all $\boldsymbol{x}, \boldsymbol{y} \in \partial C$ and $\boldsymbol{\varphi} \in \boldsymbol{S}$, that is $\mathbf{D}^R(\boldsymbol{S})$ is equicontinuous. Since $\boldsymbol{C}(\partial C)$ is dense in $\boldsymbol{L}^2(\partial C)$, the assertion follows from the Ascoli-Arzelà Theorem 3.2.3. □

As stated in the introduction of this section, the injectivity of the operator $1/2\mathbf{I} + \mathbf{K} + \mathbf{D}^R$ follows proving the uniqueness of (3.31). This is the topic of the next theorem.

Theorem 3.2.8 (Uniqueness) *The boundary value problem* (3.31) *admits a unique solution.*

Proof The proof is really similar to that one of Theorem 3.1.7. We only insert it for the reader's convenience.

Let $\boldsymbol{u}^1$ and $\boldsymbol{u}^2$ be solutions to (3.31). Then the difference $\boldsymbol{v} := \boldsymbol{u}^1 - \boldsymbol{u}^2$ solves the homogeneous version of (3.31), i.e.,

$$\begin{cases} \operatorname{div}(\mathbb{C}\widehat{\nabla}\boldsymbol{v}) = \mathbf{0} & \text{in } \mathbb{R}^3_- \setminus C \\ \dfrac{\partial \boldsymbol{v}}{\partial \boldsymbol{\nu}} = \mathbf{0} & \text{on } \partial C \\ \dfrac{\partial \boldsymbol{v}}{\partial \boldsymbol{\nu}} = \mathbf{0} & \text{on } \mathbb{R}^2 \\ \boldsymbol{v} = O(|\boldsymbol{x}|^{-1}), \qquad \nabla\boldsymbol{v} = o(|\boldsymbol{x}|^{-1}) & |\boldsymbol{x}| \to \infty, \end{cases} \tag{3.45}$$

where the decay condition on $\boldsymbol{v}$ comes from Corollary 3.2.2.

Applying Betti's formula (3.4) to $\boldsymbol{v}$ in $\Omega_r = (\mathbb{R}^3_- \cap B_r^-(\mathbf{0})) \setminus C$, we find

$$\int_{\partial\Omega_r} \boldsymbol{v} \cdot \frac{\partial \boldsymbol{v}}{\partial \boldsymbol{\nu}} \, d\sigma(\boldsymbol{x}) = \int_{\Omega_r} \mathbb{C}\widehat{\nabla}\boldsymbol{v} : \widehat{\nabla}\boldsymbol{v} \, d\boldsymbol{x}.$$

Note that $\partial\Omega_r = \partial C \cup \partial B_r^b(\mathbf{0}) \cup \partial B_r^h(\mathbf{0})$, where we recall that $\partial B_r^b(\mathbf{0})$ is the spherical cap and $\partial B_r^h(\mathbf{0})$ is the intersection of the half ball with $\{x_3 = 0\}$. From the boundary conditions in (3.45) we have that

$$\int_{\partial C} \boldsymbol{v} \cdot \frac{\partial \boldsymbol{v}}{\partial \boldsymbol{\nu}} \, d\sigma(\boldsymbol{x}) = 0, \qquad \int_{\partial B_r^h(\mathbf{0})} \boldsymbol{v} \cdot \frac{\partial \boldsymbol{v}}{\partial \boldsymbol{\nu}} \, d\sigma(\boldsymbol{x}) = 0.$$

Instead, from the decay conditions, we infer that on the spherical cap

$$\left| \int_{\partial B_r^b(\mathbf{0})} \boldsymbol{v} \cdot \frac{\partial \boldsymbol{v}}{\partial \boldsymbol{\nu}} \, d\sigma(\boldsymbol{x}) \right| \leq \int_{\partial B_r^b(\mathbf{0})} |\boldsymbol{v}| \left| \frac{\partial \boldsymbol{v}}{\partial \boldsymbol{\nu}} \right| d\sigma(\boldsymbol{x}) \leq \frac{C}{r} \int_{\partial B_r^b(\mathbf{0})} \left| \frac{\partial \boldsymbol{v}}{\partial \boldsymbol{\nu}} \right| d\sigma(\boldsymbol{x}).$$

Observe that using the decay conditions at infinity for the gradient of the function $\boldsymbol{v}$ the last integral in formula above tends to zero as $r \to +\infty$. For the proof, one can follow the same argument of (3.35) in which the spherical coordinates were applied. Therefore

$$\int_{\mathbb{R}^3_- \setminus C} \mathbb{C}\widehat{\nabla} \boldsymbol{v} : \widehat{\nabla} \boldsymbol{v} \, d\boldsymbol{x} = 0.$$

From the strong convexity of the tensor $\mathbb{C}$, see (3.3), we find that $\widehat{\nabla} \boldsymbol{v} = \mathbf{0}$ in $\mathbb{R}^3_- \setminus C$, which implies that $\boldsymbol{v} = \boldsymbol{a} + \mathbf{A}\boldsymbol{x}$, where $\boldsymbol{a} \in \mathbb{R}^3$ and $\mathbf{A} \in \mathbb{R}^{3\times3}$ is a skew matrix (see (3.12). However, we find that $\boldsymbol{a}$ and $\mathbf{A}$ must be identically null thanks to the behavior of the function $\boldsymbol{v}$ and its gradient at infinity, see (3.45). Therefore $\boldsymbol{v} = \mathbf{0}$.

□

Remark 3.2.9 From this uniqueness result, the injectivity of the operator (3.43) follows.

We recall, for the reader's convenience, the definition of the index of an operator (see [1, 99]), in order to prove the surjectivity.

Definition 3.2.1 ([1, 99]) Given a bounded operator $T : X \to Y$ between two Banach spaces, we call $\dim(\text{Ker}(T))$ and $\dim(Y/\text{Im}(T))$ the **nullity** and the **defect** of T respectively. If the nullity or the defect of T is finite, then we define the index of the operator T as the extended real number given by

$$i(T) = \dim(\text{Ker}(T)) - \dim(Y/\text{Im}(T)).$$

In addition, the operator T is said to be

1. **Semi-Fredholm** if it has a closed range and either its nullity or its defect is finite;
2. **Fredholm** if the nullity and the defect are both finite.

For our purposes we need to recall that if a bounded operator $T : X \to Y$ between Banach spaces has a finite defect then it has a closed range, see, for example, Lemma 4.38 in [1]. In addition, we also recall the following theorem on the index of a compact perturbation of a Fredholm operator which is a consequence of the theorem of F.V. Atkinson on the characterization of Fredholm operators, see [1].

Theorem 3.2.10 ([1]) *Let $T : X \to Y$ be a bounded linear operator of Fredholm type and $K : X \to Y$ a compact operator between the two Banach spaces X and Y. Then $T + K$ is Fredholm with index $i(T + K) = i(T)$.*

Finally we prove the surjectivity theorem.

Theorem 3.2.11 *The operator* $\frac{1}{2}\mathbf{I}+\mathbf{K}+\mathbf{D}^R$ *is onto in* $\boldsymbol{L}^2(\partial C)$.

Proof As a consequence of Lemma 3.2.6, the operator $\frac{1}{2}\mathbf{I}+\mathbf{K} : \boldsymbol{L}^2(\partial C) \to \boldsymbol{L}^2(\partial C)$ is Fredholm with index $i\left(\frac{1}{2}\mathbf{I}+\mathbf{K}\right) = 0$. Indeed, it is injective and surjective hence its nullity and defect are null. In addition, since $\mathbf{D}^R : \boldsymbol{L}^2(\partial C) \to \boldsymbol{L}^2(\partial C)$ is a compact operator, see Lemma 3.2.7, it follows, by the application of Theorem 3.2.10, that

$$i\left(\frac{1}{2}\mathbf{I}+\mathbf{K}+\mathbf{D}^R\right) = 0.$$

By using the definition of the index of an operator we find that

$$\dim\left(\text{Ker}\left(\frac{1}{2}\mathbf{I}+\mathbf{K}+\mathbf{D}^R\right)\right) = \dim\left(\mathbf{L}^2(\partial C)/\text{Im}\left(\frac{1}{2}\mathbf{I}+\mathbf{K}+\mathbf{D}^R\right)\right).$$

Therefore,by Remark 3.2.9, it follows that $\dim(\text{Ker}(\frac{1}{2}\mathbf{I}+\mathbf{K}+\mathbf{D}^R)) = 0$ which implies that $\dim(\mathbf{L}^2(\partial C)/\text{Im}(\frac{1}{2}\mathbf{I}+\mathbf{K}+\mathbf{D}^R)) = 0$. This means that the operator $\frac{1}{2}\mathbf{I}+\mathbf{K}+\mathbf{D}^R$ has a closed range (since its defect is null), hence

$$\text{Im}\left(\frac{1}{2}\mathbf{I}+\mathbf{K}+\mathbf{D}^R\right) = \boldsymbol{L}^2(\partial C).$$

□

Remark 3.2.12 The existence of a solution of (3.31) follows from this surjectivity result.

The well-posedness of (3.31) finally follows collecting all previous results.

Corollary 3.2.13 *There exists a unique solution to* (3.31).

Proof Uniqueness follows from Theorem 3.2.8 and the existence from Remark 3.2.12. □

3.2.2 Well-Posedness via Weighted Sobolev Spaces

As stated in the introduction of this chapter, we now present the well-posedness using a variational approach in weighted Sobolev spaces. Their definition can be found in Sect. 2.3.1, precisely in Definition 2.1.4, choosing $d = 3$. We always use the hypothesis (3.2) on the Lamé parameters. This approach is more suitable to get quantitative stability estimates of the solution to

$$\begin{cases} \operatorname{div}(\mathbb{C}\widehat{\nabla}\boldsymbol{u}) = \boldsymbol{0} & \text{in } \mathbb{R}^3_- \setminus \overline{C} \\ (\mathbb{C}\widehat{\nabla}\boldsymbol{u})\boldsymbol{n} = p\boldsymbol{n} & \text{on } \partial C \\ (\mathbb{C}\widehat{\nabla}\boldsymbol{u})\boldsymbol{e}_3 = \boldsymbol{0} & \text{on } \mathbb{R}^2, \end{cases} \tag{3.46}$$

since they are consequences of the Lax–Milgram theorem. The adjective "quantitative" means that we are able to explicit the dependence of the constants, which appear in the estimate of the solution, in terms of some a priori information on the cavity. As we will see later, stability estimates of the direct problem are used to find stability estimates for the inverse problem. For this reason, we need to list here the assumptions which we make on the cavity C.

Assumptions—Direct Problem We assume Lipschitz regularity, that is,

$$\partial C \text{ is Lipschitz with constants } r_0 \text{ and } E_0 \tag{3.47}$$

and, denoting with $\operatorname{diam}(A)$ the diameter of a set A, we require

$$B^-_{2D_0}(\boldsymbol{0}) \supset C, \tag{3.48}$$

$$d(C, \mathbb{R}^2) \geq D_0, \tag{3.49}$$

$$\operatorname{diam}(C) < D_0, \tag{3.50}$$

where the constant $D_0 > 1$ without loss of generality.

For the definition of the constants r_0 and E_0 see Definition 1.1.1. In the estimates that will follow the constants will depend on the Lamé parameters λ and μ, on r_0, E_0, and D_0.

We recall the definition of the weighted Sobolev space of our interest (see also Remark 2.1.4 for the general case)

$$H^1_w(\mathbb{R}^3_- \setminus \overline{C}) = \left\{\boldsymbol{u} \in \mathcal{D}'(\mathbb{R}^3_- \setminus \overline{C}), \frac{\boldsymbol{u}}{\rho} \in L^2(\mathbb{R}^3_- \setminus \overline{C}), \nabla \boldsymbol{u} \in L^2(\mathbb{R}^3_- \setminus \overline{C})\right\}, \tag{3.51}$$

where $\rho = (1 + |\boldsymbol{x}|^2)^{-1/2}$, see Eq. (2.3) in Sect. 2.1, with the norm given by

$$\|\boldsymbol{u}\|^2_{H^1_w(\mathbb{R}^3_- \setminus \overline{C})} = \left(\|\rho^{-1}\boldsymbol{u}\|^2_{L^2(\mathbb{R}^3_- \setminus \overline{C})} + \|\nabla \boldsymbol{u}\|^2_{L^2(\mathbb{R}^3_- \setminus \overline{C})}\right). \tag{3.52}$$

Our goal is to prove well-posedness and find quantitative stability estimates of the solution in $H^1_w(\mathbb{R}^3_- \setminus \overline{C})$, which will be used in the treatment of the inverse problem. To prove the existence and uniqueness of the solution to (3.46), essentially we need to study the continuity and coercivity of the bilinear form associated to the problem in (3.46). As we will see, coercivity comes from weighted Poincaré and Korn-type inequalities. Since we want to find quantitative estimates for the solution, we need to utilize quantitative form of these two inequalities. In [37] the authors proved a

quantitative version of these two inequalities in $H^1_w(\mathbb{R}^3_- \setminus \overline{C})$. For completeness, in the next lines we recall how to prove these results.

One way to proceed to prove a weighted Poincaré and Korn-type inequality in $H^1_w(\mathbb{R}^3_- \setminus \overline{C})$ is to use a suitable partition of the unity. For this reason, we consider two half balls, which we call $B^-_r(\mathbf{0})$ and $B^-_R(\mathbf{0})$, with $r < R$, such that

$$C \subset B^-_r(\mathbf{0}) \subset B^-_R(\mathbf{0}).$$

Using the a priori information (3.49) and (3.50) on the cavity C, we can, for instance, fix

$$r = 3D_0 \qquad \text{and} \qquad R = 4D_0.$$

We are now in the position to construct a specific partition of the unity of $\mathbb{R}^3_-$. In particular, we take $\varphi_1, \varphi_2 \in C^\infty(\mathbb{R}^3_-)$ such that

$$0 \le \varphi_1, \varphi_2 \le 1 \quad \text{and} \quad \varphi_1 + \varphi_2 = 1 \quad \text{in } \mathbb{R}^3_-, \tag{3.53}$$

with

$$\varphi_2 = 0, \quad \varphi_1 = 1, \qquad \text{in } B^-_r(\mathbf{0}), \tag{3.54}$$

$$\varphi_1 = 0, \quad \varphi_2 = 1, \qquad \text{in } \{|\boldsymbol{x}| \ge R\} \cap \mathbb{R}^3_-, \tag{3.55}$$

$$|\nabla \varphi_1| \le \frac{c}{\rho}, \quad |\nabla \varphi_2| \le \frac{c}{\rho}, \qquad \text{in } B^-_R(\mathbf{0}) \setminus B^-_r(\mathbf{0}). \tag{3.56}$$

It is straightforward to see that, for the choice made for r and R, the constant c is an absolute positive constant, i.e., does not depend on D_0.

For the following proof, it is useful to recall and adapt a generalization of Hardy's inequality which is contained in [102], see Lemma 3 p. 83.

Proposition 3.2.14 *Let* $\zeta > 0$. *For any* $\boldsymbol{u} \in H^1_w(\mathbb{R}^3_- \cap \{|\boldsymbol{x}| \ge \zeta\})$ *it holds*

$$\int_{\mathbb{R}^3_- \cap \{|\boldsymbol{x}| \ge \zeta\}} \left|\frac{\boldsymbol{u}}{\rho}\right|^2 d\boldsymbol{x} \le C \int_{\mathbb{R}^3_- \cap \{|\boldsymbol{x}| \ge \zeta\}} |\nabla \boldsymbol{u}|^2 \, d\boldsymbol{x},$$

where the constant C *does not depend on* $\boldsymbol{u}$.

Proof The proof is a small variation of that one contained in [102] of Lemma 2 p. 82 and Lemma 3 p. 83. For this reason we only put in evidence changes to do. We note that the statement of the theorem can be proved componentwise for the displacement vector, i.e.,

$$\int_{\mathbb{R}^3_- \cap \{|\boldsymbol{x}| \ge \zeta\}} \frac{u_i^2}{\rho^2} \, d\boldsymbol{x} \le C \int_{\mathbb{R}^3_- \cap \{|\boldsymbol{x}| \ge \zeta\}} |\nabla u_i|^2 \, d\boldsymbol{x}, \qquad \text{for } i = 1, 2, 3.$$

For our purposes, it is sufficient to take $a = 0$, $n = 3$, and $p = 2$, from the statement of Lemma 2 in [102]. We define the function $\psi_i(r) := \int_{K_1} u_i(r, \boldsymbol{\theta})\, dS_1$, where $(r, \boldsymbol{\theta})$ are the polar coordinates with $r \geq \zeta$ and $K_1 = \mathbb{R}^3_- \cap \{|\boldsymbol{x}| = 1\}$. By the application of the Hölder inequality, we get

$$\int_\zeta^{+\infty} r^2 (\psi_i')^2\, dr = \int_\zeta^{+\infty} r^2 \left(\int_{K_1} \frac{\partial u_i}{\partial r}\, dS_1 \right)^2 dr \leq C_1 \int_{\mathbb{R}^3_- \cap \{|\boldsymbol{x}| \geq \zeta\}} |\nabla u_i|^2\, d\boldsymbol{x}. \tag{3.57}$$

This inequality implies that there exists a constant $M \geq 0$ such that $\psi_i(r) \to M$, as $r \to +\infty$. Since $u_i \in H^1_w(\mathbb{R}^3_- \cap \{|\boldsymbol{x}| \geq \zeta\})$, this implies that $M = 0$. In fact,

$$|\psi_i(r_2) - \psi_i(r_1)| = \left| \int_{r_1}^{r_2} \psi_i'\, dr \right| \leq \left| \int_{r_1}^{r_2} (\psi_i')^2 r^2\, dr \right|^{1/2} \left| \int_{r_1}^{r_2} \frac{1}{r^2}\, dr \right|^{1/2}. \tag{3.58}$$

Observing that the second integral on the right-hand side tends to zero as $r_1, r_2 \to +\infty$ and the first one is bounded thanks to (3.57), we get that the left-hand side of (3.58) tends to zero as $r_1, r_2 \to +\infty$. Now, using a weighted Hardy's inequality, see, for example, [102, 109], we find that

$$\int_\zeta^{+\infty} (\psi_i)^2\, dr \leq C \int_\zeta^{+\infty} r^2 (\psi_i')^2\, dr \leq C \int_{\mathbb{R}^3_- \backslash \{|\boldsymbol{x}| \geq \zeta\}} |\nabla u_i|^2\, d\boldsymbol{x}, \tag{3.59}$$

where the constant C does not depend on u_i. Thanks to the Poincaré inequality, see [102], we have

$$\int_{K_1} |u_i - \psi_i|^2\, dS_1 \leq C \int_{K_1} r^2 |\nabla u_i|^2\, dS_1,$$

where C does not depend on r and u_i. Hence it follows that

$$\int_{K_1} |u_i|^2 dS_1 \leq C \left(\int_{K_1} r^2 |\nabla u_i|^2\, dS_1 + \int_{K_1} |\psi_i|^2\, dS_1 \right).$$

Integrating with respect to r and then passing to Cartesian coordinates, we find

$$\begin{aligned} \int_{\mathbb{R}^3_- \backslash \{|\boldsymbol{x}| \geq \zeta\}} \left(\frac{u_i}{\rho} \right)^2 d\boldsymbol{x} &\leq \int_{\mathbb{R}^3_- \backslash \{|\boldsymbol{x}| \geq \zeta\}} (u_i)^2 |\boldsymbol{x}|^{-2}\, d\boldsymbol{x} \\ &\leq C \left(\int_{\mathbb{R}^3_- \backslash \{|\boldsymbol{x}| \geq \zeta\}} |\nabla u_i|^2\, d\boldsymbol{x} + \int_{\mathbb{R}^3_- \backslash \{|\boldsymbol{x}| \geq \zeta\}} |\boldsymbol{x}|^{-2} (\psi_i)^2\, d\boldsymbol{x} \right). \end{aligned}$$

Through (3.59) the assertion of the proposition follows. □

We are now in the position to prove the following theorem.

Theorem 3.2.15 (Weighted Poincaré Inequality) *Given* ρ *in* (2.3) *and for any function* $\boldsymbol{u} \in H^1_w(\mathbb{R}^3_- \setminus \overline{C})$, *there exists a positive constant* c, *with* $c = c(r_0, E_0, D_0)$, *such that*

$$\int_{\mathbb{R}^3_- \setminus \overline{C}} \left| \frac{\boldsymbol{u}}{\rho} \right|^2 d\boldsymbol{x} \leq c \int_{\mathbb{R}^3_- \setminus \overline{C}} |\nabla \boldsymbol{u}|^2 \, d\boldsymbol{x}. \tag{3.60}$$

Proof From the property of φ, see (3.53), we first observe that

$$\left\| \frac{\boldsymbol{u}}{\rho} \right\|^2_{L^2(\mathbb{R}^3_- \setminus \overline{C})} \leq 2 \left(\left\| \varphi_1 \frac{\boldsymbol{u}}{\rho} \right\|^2_{L^2(\mathbb{R}^3_- \setminus \overline{C})} + \left\| \varphi_2 \frac{\boldsymbol{u}}{\rho} \right\|^2_{L^2(\mathbb{R}^3_- \setminus \overline{C})} \right) := 2(\mathcal{P}_1 + \mathcal{P}_2).$$

We need to find estimates of $\mathcal{P}_1$ and $\mathcal{P}_2$.

Since $\rho^{-1} \leq 1$, from the property (3.55) we find

$$\mathcal{P}_1 = \left\| \varphi_1 \frac{\boldsymbol{u}}{\rho} \right\|^2_{L^2(B^-_R(\mathbf{0}) \setminus \overline{C})} \leq \|\varphi_1 \boldsymbol{u}\|^2_{L^2(B^-_R(\mathbf{0}) \setminus \overline{C})}. \tag{3.61}$$

Observing that $\varphi_1 = 0$ on $\partial B^b_R(\mathbf{0})$, we can estimate the right-hand side of (3.61) by using a quantitative Poincaré inequality for functions vanishing on a portion of the boundary of a bounded domain. This result has been proved, for instance, in [12] (Theorem 3.3 and Example 3.6). Then

$$\|\varphi_1 \boldsymbol{u}\|^2_{L^2(B^-_R(\mathbf{0}) \setminus \overline{C})} \leq c \, \|\nabla(\varphi_1 \boldsymbol{u})\|^2_{L^2(B^-_R(\mathbf{0}) \setminus \overline{C})}, \tag{3.62}$$

where c is a positive constant such that $c = c(r_0, E_0, D_0)$. Therefore, the right-hand side of the previous inequality gives

$$\|\nabla(\varphi_1 \boldsymbol{u})\|^2_{L^2(B^-_R(\mathbf{0}) \setminus \overline{C})} \leq 2 \left(\|\boldsymbol{u} \otimes \nabla \varphi_1\|^2_{L^2(B^-_R(\mathbf{0}) \setminus \overline{C})} + \|\varphi_1 \nabla \boldsymbol{u}\|^2_{L^2(B^-_R(\mathbf{0}) \setminus \overline{C})} \right). \tag{3.63}$$

Now, utilizing the property (3.56) on the gradient of φ_1, we find

$$\|\boldsymbol{u} \otimes \nabla \varphi_1\|^2_{L^2(B^-_R(\mathbf{0}) \setminus \overline{C})} = \int_{B^-_R(\mathbf{0}) \setminus \overline{C}} |\boldsymbol{u}|^2 |\nabla \varphi_1|^2 \, d\boldsymbol{x} \leq c \int_{B^-_R(\mathbf{0}) \setminus \overline{B^-_r(\mathbf{0})}} \frac{|\boldsymbol{u}|^2}{\rho^2} \, d\boldsymbol{x}. \tag{3.64}$$

Inserting (3.64) in (3.63) and then (3.63) in (3.62) and recalling (3.53), we have

$$\|\varphi_1 \boldsymbol{u}\|^2_{L^2(B^-_R(\mathbf{0}) \setminus \overline{C})} \leq c \left(\left\| \frac{\boldsymbol{u}}{\rho} \right\|^2_{L^2(\{|\boldsymbol{x}| > r\} \cap \mathbb{R}^3_-)} + \|\nabla \boldsymbol{u}\|^2_{L^2(B^-_R(\mathbf{0}) \setminus \overline{C})} \right). \tag{3.65}$$

For the first term in the right-hand side of the inequality (3.65), we apply Hardy's inequality, see Proposition 3.2.14, that is

$$\left\| \frac{\boldsymbol{u}}{\rho} \right\|^2_{L^2(\{|\boldsymbol{x}|>r\}\cap\mathbb{R}^3_-)} \leq c\,\|\nabla \boldsymbol{u}\|^2_{L^2(\{|\boldsymbol{x}|>r\}\cap\mathbb{R}^3_-)}, \tag{3.66}$$

where the constant c is independent from $\boldsymbol{u}$. Inserting (3.66) in (3.65) and then going back to (3.61), we have

$$\begin{aligned} \mathcal{P}_1 = \left\| \varphi_1 \frac{\boldsymbol{u}}{\rho} \right\|^2_{L^2(B^-_R(\mathbf{0})\setminus\overline{C})} &\leq c\left(\|\nabla \boldsymbol{u}\|^2_{L^2(B^-_R(\mathbf{0})\setminus\overline{C})} + \|\nabla \boldsymbol{u}\|^2_{L^2(\{|\boldsymbol{x}|>r\}\cap\mathbb{R}^3_-)} \right) \\ &\leq c\,\|\nabla \boldsymbol{u}\|^2_{L^2(\mathbb{R}^3_-\setminus\overline{C})}. \end{aligned} \tag{3.67}$$

For the term $\mathcal{P}_2$, we use the properties (3.53) and (3.54) and apply again Hardy's inequality, see Proposition 3.2.14, i.e.,

$$\begin{aligned} \mathcal{P}_2 = \left\| \varphi_2 \frac{\boldsymbol{u}}{\rho} \right\|^2_{L^2(\mathbb{R}^3_-\setminus\overline{C})} \leq \left\| \frac{\boldsymbol{u}}{\rho} \right\|^2_{L^2(\{|\boldsymbol{x}|>r\}\cap\mathbb{R}^3_-)} &\leq c\|\nabla \boldsymbol{u}\|^2_{L^2(\{|\boldsymbol{x}|>r\}\cap\mathbb{R}^3_-)} \\ &\leq c\,\|\nabla \boldsymbol{u}\|^2_{L^2(\mathbb{R}^3_-\setminus\overline{C})}. \end{aligned} \tag{3.68}$$

Putting together the inequalities (3.67) and (3.68) we have the assertion. □

To prove a Korn-type inequality in the exterior domain of a half-space, we need a preliminary result about the bound of the gradient of a function $\boldsymbol{u} \in H^1_w(\mathbb{R}^3_- \setminus \overline{C})$ with its symmetric part. It is an adaptation of a lemma proved by Kondrat'ev and Oleinik in [102] (see Lemma 5, p. 85) in full generality to the case of functions $\boldsymbol{u} \in H^1_w(\mathbb{R}^3_- \setminus \overline{C})$. The only difference between our and Kondrat'ev and Oleinik result is the fact that our estimate does not contain rigid motions because they do not belong to $H^1_w(\mathbb{R}^3_- \setminus \overline{C})$.

Lemma 3.2.16 *Let $\boldsymbol{u} \in H^1_w(\mathbb{R}^3_- \setminus \overline{C})$. For every $r' < r$ there exists a positive constant c such that*

$$\|\nabla \boldsymbol{u}\|_{L^2(\{|\boldsymbol{x}|>r\}\cap\mathbb{R}^3_-)} \leq c\|\widehat{\nabla} \boldsymbol{u}\|_{L^2(\{|\boldsymbol{x}|>r'\}\cap\mathbb{R}^3_-)},$$

where $c = c(r, r')$.

Now, it follows this quantitative version of the Korn inequality.

Theorem 3.2.17 (Korn-Type Inequality) *For any function $\boldsymbol{u} \in H^1_w(\mathbb{R}^3_- \setminus \overline{C})$ there exists a positive constant c, with $c = c(r_0, E_0, D_0)$, such that*

$$\int_{\mathbb{R}^3_- \setminus \overline{C}} |\nabla \boldsymbol{u}|^2 \, dx \le c \int_{\mathbb{R}^3_- \setminus \overline{C}} |\widehat{\nabla} \boldsymbol{u}|^2 \, dx. \tag{3.69}$$

Proof Using the definition of the functions φ_1 and φ_2, see (3.53), we can estimate the gradient of the function $\boldsymbol{u}$ in the following way:

$$\|\nabla \boldsymbol{u}\|^2_{L^2(\mathbb{R}^3_- \setminus \overline{C})} \le 2\Big(\|\nabla(\varphi_1 \boldsymbol{u})\|^2_{L^2(\mathbb{R}^3_- \setminus \overline{C})} + \|\nabla(\varphi_2 \boldsymbol{u})\|^2_{L^2(\mathbb{R}^3_- \setminus \overline{C})} \Big) := 2(\mathcal{K}'_1 + \mathcal{K}'_2).$$

We estimate, separately, the two terms $\mathcal{K}'_1$ and $\mathcal{K}'_2$.

By (3.55) we find

$$\mathcal{K}'_1 = \|\nabla(\varphi_1 \boldsymbol{u})\|^2_{L^2(\mathbb{R}^3_- \setminus \overline{C})} = \|\nabla(\varphi_1 \boldsymbol{u})\|^2_{L^2(B^-_R(\mathbf{0}) \setminus \overline{C})}. \tag{3.70}$$

Since $\varphi_1 = 0$ on $\partial B^b_R(\mathbf{0})$ and we are working in a bounded domain, we apply the quantitative Korn inequality for functions vanishing on a portion of the boundary of a bounded domain, see for instance [12] (Theorem 5.7), from which it follows, for $c = c(r_0, E_0, D_0)$,

$$\begin{aligned} \|\nabla(\varphi_1 \boldsymbol{u})\|^2_{L^2(B^-_R(\mathbf{0}) \setminus \overline{C})} &\le c \, \|\widehat{\nabla}(\varphi_1 \boldsymbol{u})\|^2_{L^2(B^-_R(\mathbf{0}) \setminus \overline{C})} \\ &\le c \Big(\|\boldsymbol{u} \otimes \nabla \varphi_1\|^2_{L^2(B^-_R(\mathbf{0}) \setminus \overline{C})} + \|\varphi_1 \widehat{\nabla} \boldsymbol{u}\|^2_{L^2(B^-_R(\mathbf{0}) \setminus \overline{C})} \Big), \end{aligned}$$

where for the first term in the right-hand side of the previous inequality we have used

$$\|\widehat{\boldsymbol{u} \otimes \nabla \varphi_1}\|^2_{L^2(B^-_R(\mathbf{0}) \setminus \overline{C})} \le \|\boldsymbol{u} \otimes \nabla \varphi_1\|^2_{L^2(B^-_R(\mathbf{0}) \setminus \overline{C})}.$$

Now, utilizing (3.64), the properties (3.53) and Hardy's inequality (3.66), we have

$$\begin{aligned} \|\boldsymbol{u} \otimes \nabla \varphi_1\|^2_{L^2(B^-_R(\mathbf{0}) \setminus \overline{C})} &+ \|\varphi_1 \widehat{\nabla} \boldsymbol{u}\|^2_{L^2(B^-_R(\mathbf{0}) \setminus \overline{C})} \\ &\le c \left(\left\| \frac{\boldsymbol{u}}{\rho} \right\|^2_{L^2(\{|\boldsymbol{x}|>r\} \cap \mathbb{R}^3_-)} + \|\widehat{\nabla} \boldsymbol{u}\|^2_{L^2(B^-_R(\mathbf{0}) \setminus \overline{C})} \right) \\ &\le c \left(\|\nabla \boldsymbol{u}\|^2_{L^2(\{|\boldsymbol{x}|>r\} \cap \mathbb{R}^3_-)} + \|\widehat{\nabla} \boldsymbol{u}\|^2_{L^2(B^-_R(\mathbf{0}) \setminus \overline{C})} \right). \end{aligned}$$

Applying Lemma 3.2.16 to the first term in the right-hand side of the previous formula, we find

$$\|\nabla \boldsymbol{u}\|^2_{L^2(\{|\boldsymbol{x}|>r\} \cap \mathbb{R}^3_-)} \le c \|\widehat{\nabla} \boldsymbol{u}\|^2_{L^2(\{|\boldsymbol{x}|>r'\} \cap \mathbb{R}^3_-)},$$

where we choose $r' = 2D_0 < r$ so that $C \subset B^-_{r'}(\mathbf{0})$ (see the a priori information (3.48)). From all these results and going back to (3.70), we find

$$\mathcal{K}'_1 \leq c\left(\|\widehat{\nabla}\boldsymbol{u}\|^2_{L^2(B^-_R(\mathbf{0})\setminus\overline{C})} + \|\widehat{\nabla}\boldsymbol{u}\|^2_{L^2(\{|\boldsymbol{x}|>r'\}\cap\mathbb{R}^3_-)}\right) \leq c\,\|\widehat{\nabla}\boldsymbol{u}\|^2_{L^2(\mathbb{R}^3_-\setminus\overline{C})}. \tag{3.71}$$

In a similar way, using the properties (3.53) and (3.54), we find

$$\begin{aligned}\mathcal{K}'_2 &= \|\nabla(\varphi_2\boldsymbol{u})\|^2_{L^2(\mathbb{R}^3_-\setminus\overline{C})} = \|\nabla(\varphi_2\boldsymbol{u})\|^2_{L^2(\{|\boldsymbol{x}|>r\}\cap\mathbb{R}^3_-)}\\ &\leq c\left(\|\boldsymbol{u}\otimes\nabla\varphi_2\|^2_{L^2(\{|\boldsymbol{x}|>r\}\cap\mathbb{R}^3_-)} + \|\varphi_2\nabla\boldsymbol{u}\|^2_{L^2(\{|\boldsymbol{x}|>r\}\cap\mathbb{R}^3_-)}\right).\end{aligned}$$

Utilizing properties (3.53) and (3.56) of φ_2, we get

$$\begin{aligned}&\|\boldsymbol{u}\otimes\nabla\varphi_2\|^2_{L^2(\{|\boldsymbol{x}|>r\}\cap\mathbb{R}^3_-)} + \|\varphi_2\nabla\boldsymbol{u}\|^2_{L^2(\{|\boldsymbol{x}|>r\}\cap\mathbb{R}^3_-)}\\ &\qquad\leq c\left(\left\|\frac{\boldsymbol{u}}{\rho}\right\|^2_{L^2(\{|\boldsymbol{x}|>r\}\cap\mathbb{R}^3_-)} + \|\nabla\boldsymbol{u}\|^2_{L^2(\{|\boldsymbol{x}|>r\}\cap\mathbb{R}^3_-)}\right).\end{aligned}$$

Finally, for the last two terms of the previous formula, we apply the Hardy's inequality (3.66) and the result in Lemma 3.2.16 from which

$$\mathcal{K}'_2 \leq c\,\|\widehat{\nabla}\boldsymbol{u}\|^2_{L^2(\{|\boldsymbol{x}|>r'\}\cap\mathbb{R}^3_-)} \leq c\,\|\widehat{\nabla}\boldsymbol{u}\|^2_{L^2(\mathbb{R}^3_-\setminus\overline{C})}. \tag{3.72}$$

The assertion of the statement follows by collecting the results in (3.71) and (3.72). □

Well-Posedness

We are now in position to prove that there exists a unique weak solution of the elastic problem (3.46) in $H^1_w(\mathbb{R}^3_-\setminus\overline{C})$.

We first write a weak formulation of the problem. To do that we assume, for the moment, $\boldsymbol{u}$ regular and the test functions $\boldsymbol{v}$ in $\mathcal{D}(\overline{\mathbb{R}}^3_-\setminus C)$. We first multiply the equation $\operatorname{div}(\mathbb{C}\widehat{\nabla}\boldsymbol{u})$ in (3.46) for $\boldsymbol{v}$. Then we integrate in $\mathbb{R}^3_-\setminus\overline{C}$ by parts using boundary conditions, hence it follows that

$$\int_{\mathbb{R}^3_-\setminus\overline{C}} \mathbb{C}\widehat{\nabla}\boldsymbol{u} : \widehat{\nabla}\boldsymbol{v}\,d\boldsymbol{x} = -p\int_{\partial C} \boldsymbol{n}\cdot\boldsymbol{v}\,d\sigma(\boldsymbol{x}), \quad \forall\boldsymbol{v}\in\mathcal{D}(\overline{\mathbb{R}}^3_-\setminus C).$$

This formulation continues to be true in $H^1_w(\mathbb{R}^3_- \setminus \overline{C})$ since the functional space $\mathcal{D}(\overline{\mathbb{R}}^3_- \setminus C)$ is dense onto the weighted Sobolev space $H^1_w(\mathbb{R}^3_- \setminus \overline{C})$, see Sect. 2.1.1. Therefore problem (3.46) becomes:

Find $\boldsymbol{u} \in H^1_w(\mathbb{R}^3_- \setminus \overline{C})$ *such that*

$$a(\boldsymbol{u}, \boldsymbol{v}) = f(\boldsymbol{v}), \quad \forall \boldsymbol{v} \in H^1_w(\mathbb{R}^3_- \setminus \overline{C}), \tag{3.73}$$

where $a : H^1_w(\mathbb{R}^3_- \setminus \overline{C}) \times H^1_w(\mathbb{R}^3_- \setminus \overline{C}) \to \mathbb{R}$ *is the bilinear form given by*

$$a(\boldsymbol{u}, \boldsymbol{v}) = \int\limits_{\mathbb{R}^3_- \setminus \overline{C}} \mathbb{C}\widehat{\nabla}\boldsymbol{u} : \widehat{\nabla}\boldsymbol{v}\, d\boldsymbol{x}, \tag{3.74}$$

and $f : H^1_w(\mathbb{R}^3_- \setminus \overline{C}) \to \mathbb{R}$ *is the linear functional given by*

$$f(\boldsymbol{v}) = -p \int\limits_{\partial C} \boldsymbol{n} \cdot \boldsymbol{v}\, d\sigma(\boldsymbol{x}). \tag{3.75}$$

We finally have

Theorem 3.2.18 *Problem* (3.46) *admits a unique solution* $\boldsymbol{u} \in H^1_w(\mathbb{R}^3_- \setminus \overline{C})$ *satisfying*

$$\|\boldsymbol{u}\|_{H^1_w(\mathbb{R}^3_- \setminus \overline{C})} \leq cp, \tag{3.76}$$

where the constant $c = c(\lambda, \mu, r_0, E_0, D_0)$.

Proof Well-posedness of problem (3.46) follows by the application of the Lax–Milgram theorem to (3.73). This means that we need to prove coercivity and continuity of the bilinear form (3.74) and continuity of the linear functional (3.75).

Continuity and Coercivity of (3.74)
From the Cauchy–Schwarz inequality we have

$$\begin{aligned}|a(\boldsymbol{u}, \boldsymbol{v})| = \left| \int\limits_{\mathbb{R}^3_- \setminus \overline{C}} \mathbb{C}\widehat{\nabla}\boldsymbol{u} : \widehat{\nabla}\boldsymbol{v}\, d\boldsymbol{x} \right| &\leq c\, \|\widehat{\nabla}\boldsymbol{u}\|_{L^2(\mathbb{R}^3_- \setminus \overline{C})} \|\widehat{\nabla}\boldsymbol{v}\|_{L^2(\mathbb{R}^3_- \setminus \overline{C})} \\ &\leq c\, \|\boldsymbol{u}\|_{H^1_w(\mathbb{R}^3_- \setminus \overline{C})} \|\boldsymbol{v}\|_{H^1_w(\mathbb{R}^3_- \setminus \overline{C})},\end{aligned}$$

where $c = c(\lambda, \mu)$. Moreover, the application of the quantitative version of the Poincaré and Korn inequalities, proved in Theorems 3.2.15 and 3.2.17, respectively, and the strong convexity condition of $\mathbb{C}$, see (3.3), give the coercivity of the bilinear form (3.74). Indeed

$$a(\boldsymbol{u},\boldsymbol{u}) = \int\limits_{\mathbb{R}^3_-\setminus\overline{C}} \mathbb{C}\widehat{\nabla}\boldsymbol{u} : \widehat{\nabla}\boldsymbol{u}\, dx \geq c\|\widehat{\nabla}\boldsymbol{u}\|^2_{L^2(\mathbb{R}^3_-\setminus\overline{C})}$$

$$\geq c\|\nabla\boldsymbol{u}\|^2_{L^2(\mathbb{R}^3_-\setminus\overline{C})} \geq c\|\boldsymbol{u}\|^2_{H^1_w(\mathbb{R}^3_-\setminus\overline{C})},$$

where the constant $c = c(\lambda, \mu, r_0, E_0, D_0)$.

Continuity of (2.38)
Let us take $B^-_{2D_0}(\mathbf{0})$. The result follows by the application of the trace theorem for bounded domains, see Theorem 2.1.1, indeed

$$\left| - p \int\limits_{\partial C} \boldsymbol{n}\cdot\boldsymbol{v}\, d\sigma(\boldsymbol{x}) \right| \leq c\, p\|\boldsymbol{v}\|_{L^2(\partial C)}$$

$$\leq c\, p \left(\left\| \frac{\boldsymbol{v}}{\rho} \right\|_{L^2((B^-_{2D_0}(\mathbf{0}))\setminus\overline{C})} + \|\nabla\boldsymbol{v}\|_{L^2((B^-_{2D_0}(\mathbf{0}))\setminus\overline{C})} \right)$$

$$\leq c\, p\|\boldsymbol{v}\|_{H^1_w(\mathbb{R}^3_-\setminus\overline{C})}.$$

Thanks to the previous results, the well-posedness of (3.46) follows by Lax–Milgram. Finally, we find quantitative stability estimates for the solution of the direct problem employing again Poincaré and Korn inequalities, see Theorems 3.2.15 and 3.2.17, and the strong convexity of $\mathbb{C}$, see (3.3). In fact,

$$\|\boldsymbol{u}\|^2_{H^1_w(\mathbb{R}^3_-\setminus C)} \leq \left| \int\limits_{\mathbb{R}^3_-\setminus\overline{C}} \mathbb{C}\widehat{\nabla}\boldsymbol{u} : \widehat{\nabla}\boldsymbol{u}\, dx \right| \leq \left| p \int\limits_{\partial C} \boldsymbol{n}\cdot\boldsymbol{u}\, d\sigma(\boldsymbol{x}) \right| \leq cp\|\boldsymbol{u}\|_{H^1_w(\mathbb{R}^3_-\setminus\overline{C})},$$

where the constant $c = c(\lambda, \mu, r_0, E_0, D_0)$, hence the assertion of the theorem follows. □

3.3 The Inverse Problem: Stability Estimates

In this section we study in detail the inverse problem presented in Chap. 1, which corresponds to answer, from a mathematical point of view, to this question: *Given the displacement vector* $\boldsymbol{u}$ *on a portion of the boundary of the half-space can we detect uniquely and in a stable way the cavity* C?

As explained in Chap. 1, the main tool which is utilized to finding stability estimates for this kind of inverse problems consists, essentially, in quantitative estimates of unique continuation: stability estimates of continuation from Cauchy data (see, for instance, [90]), from the interior [8], three-spheres inequalities [2, 4, 10, 105] and doubling inequalities in the interior [13, 86]. One can consult

[7, 115], where these techniques were employed to find stability estimates for the inverse problem of detection of cavities for scalar elliptic equations and the Lamé system, respectively. For applications of these techniques in other contexts arising in inverse problems see, for example, [5, 8–11].

In the next lines, we set the assumptions which are utilized to solve our inverse problems. Then, in the following subsections, we prove the results we need to find the stability estimates.

In addition to the a priori assumptions (3.2), (3.48)–(3.50), we require

Assumptions—Inverse Problem We assume that the measurements of the displacement field are given on

$$B'_{s_0}(\mathbf{0}) = \{\boldsymbol{x} \in \mathbb{R}^2 \,:\, x_1^2 + x_2^2 < s_0^2\}, \tag{3.77}$$

which is contained in $\{x_3 = 0\}$, with $s_0 < D_0$.

To obtain stability estimates for the inverse problem we need to require more regularity on C than the Lipschitz one. Specifically, we assume that

$$\partial C \text{ is of class } C^3 \text{ with constant } r_0 \text{ and } E_0. \tag{3.78}$$

In addition, we recall that C satisfy the a priori information (3.48)–(3.50). We also assume that

$$\mathbb{R}^3_- \setminus \overline{C} \text{ is connected.}$$

If C_1 and C_2 are two cavities, we define

$$G \text{ the unbounded connected component of } \mathbb{R}^3_- \setminus (\overline{C_1 \cup C_2}). \tag{3.79}$$

It is straightforward to note that $B'_{s_0}(\mathbf{0}) \subset \partial G$.

Finally, we assume that

$$\text{the pressure } p \text{ acting on } \partial C \text{ is known.} \tag{3.80}$$

Utilizing the hypothesis (3.80), we introduce the displacement field

$$\overline{\boldsymbol{u}} = \frac{p}{3\lambda + 2\mu}\boldsymbol{x}, \tag{3.81}$$

which is helpful to reduce the direct problem (3.46) to a problem with homogeneous Neumann boundary conditions on the boundary of the cavity. It is straightforward to check that $\overline{\boldsymbol{u}}$ satisfies the Lamé system and the same boundary condition on C satisfied by $\boldsymbol{u}$. By means of this auxiliary displacement field, we can define

$$\boldsymbol{w} := \boldsymbol{u} - \overline{\boldsymbol{u}}, \tag{3.82}$$

which satisfies the following boundary value problem

$$\begin{cases} \operatorname{div}(\mathbb{C}\widehat{\nabla}\boldsymbol{w}) = \boldsymbol{0} & \text{in } \mathbb{R}^3_- \setminus \overline{C} \\ (\mathbb{C}\widehat{\nabla}\boldsymbol{w})\boldsymbol{n} = 0 & \text{on } \partial C \\ (\mathbb{C}\widehat{\nabla}\boldsymbol{w})\boldsymbol{e}_3 = -p\boldsymbol{e}_3 & \text{on } \mathbb{R}^2 \\ \boldsymbol{w} + \overline{\boldsymbol{u}} \in H^1_w(\mathbb{R}^3_- \setminus \overline{C}), \end{cases} \tag{3.83}$$

where $\boldsymbol{e}_3 = (0, 0, 1)$. We can now state precisely our inverse problem.

The Inverse Problem Determine the cavity C from a single pair of the Cauchy data on $B'_{s_0}(\boldsymbol{0})$ of the solution to (3.83).

Given two cavities C_i, for $i = 1, 2$, we denote by $\boldsymbol{w}_i$, for $i = 1, 2$, the difference

$$\boldsymbol{w}_i = \boldsymbol{u}_i - \overline{\boldsymbol{u}}, \qquad \text{for } i = 1, 2, \tag{3.84}$$

where $\boldsymbol{w}_i$ and $\boldsymbol{u}_i$ are respectively the solutions to (3.83) and (3.46) with $C = C_i$, for $i = 1, 2$. From the definition of the function $\boldsymbol{w}$, it immediately follows that

$$\boldsymbol{w}_1 - \boldsymbol{w}_2 = \boldsymbol{u}_1 - \boldsymbol{u}_2, \qquad \text{in } \mathbb{R}^3_- \setminus \overline{(C_1 \cup C_2)}. \tag{3.85}$$

3.3.1 A Regularity Result

Utilizing the a priori assumption (3.78), we can prove that the solution $\boldsymbol{u}$ of the direct problem (3.46) is more regular. For this purpose, we adopt the integral representation formula given in (3.33) with (3.34), and we consider a bounded domain $Q \subset \mathbb{R}^3_-$ such that

$$\partial Q \in C^3 \text{ with constants } r_0, E_0, \tag{3.86}$$

$$\overline{B^-_{\alpha D_0}(\boldsymbol{0})} \subset\subset \overline{Q} \subset\subset \overline{B^-_{\beta D_0}(\boldsymbol{0})}, \tag{3.87}$$

where $\alpha > 2$ and $\beta \geq 3$, with $\alpha < \beta$, and we recall that for this choice of α the cavity $C \subset B^-_{\alpha D_0}(\boldsymbol{0})$, see Assumptions (3.48). It holds

Proposition 3.3.1 *Under the assumptions* (3.78) *for C and* (3.86), (3.87) *for Q, the solution of problem* (3.46), *satisfies*

$$\|\boldsymbol{u}\|_{C^{1,1/2}(\overline{Q}\setminus C)} \leq cp, \tag{3.88}$$

where the constant $c = c(\lambda, \mu, \alpha, \beta, r_0, E_0, D_0)$.

Proof of Theorem 3.3.1 Taking $\boldsymbol{y} \in \partial Q \setminus \partial C$, the kernels of the integral operators in (3.33) are regular, see also the explicit expression of the Neumann function in Theorem 3.1.7. Then we can estimate $D^k\boldsymbol{u}(\boldsymbol{y})$, for $k = 0, 1, 2, 3$, that is

$$|D^k\boldsymbol{u}(\boldsymbol{y})| \le p|\partial C| \sup_{\substack{\boldsymbol{x}\in\partial C\\ \boldsymbol{y}\in\partial Q\setminus\partial C}} |D^k_{\boldsymbol{y}}\mathbf{N}(\boldsymbol{x},\boldsymbol{y})| + |\partial C|^{1/2}\|\boldsymbol{f}\|_{L^2(\partial C)} \sup_{\substack{\boldsymbol{x}\in\partial C\\ \boldsymbol{y}\in\partial Q\setminus\partial C}} |D^k_{\boldsymbol{y}}(\mathbb{C}\widehat{\nabla}_{\boldsymbol{x}}\mathbf{N}(\boldsymbol{x},\boldsymbol{y}))|.$$

Utilizing the regularity properties of $\mathbf{N}$ and Theorem 3.1.7 (in particular properties (3.15) and its generalizations), we yield

$$\sup_{\substack{\boldsymbol{x}\in\partial C\\ \boldsymbol{y}\in\partial Q\setminus\partial C}} |D^k_{\boldsymbol{y}}\mathbf{N}(\boldsymbol{x},\boldsymbol{y})| \le \frac{c}{D_0^{k+1}}, \qquad \sup_{\substack{\boldsymbol{x}\in\partial C\\ \boldsymbol{y}\in\partial Q\setminus\partial C}} |D^k_{\boldsymbol{y}}(\mathbb{C}\widehat{\nabla}_{\boldsymbol{x}}\mathbf{N}(\boldsymbol{x},\boldsymbol{y}))| \le \frac{c}{D_0^{k+2}}, \tag{3.89}$$

where the constant $c = c(\lambda, \mu, \alpha)$. From the trace estimate applied in Q, we have

$$\|\boldsymbol{f}\|_{L^2(\partial C)} \le c\|\boldsymbol{u}\|_{H^1_w(\mathbb{R}^3_-\setminus\overline{C})} \le cp, \tag{3.90}$$

hence, from (3.89) and (3.90), we get

$$|D^k\boldsymbol{u}(\boldsymbol{y})| \le cp,$$

where the constant $c = c(\lambda, \mu, \alpha, r_0, E_0, D_0)$. Therefore

$$\|D^k\boldsymbol{u}\|_{L^\infty(\partial Q\setminus\partial C)} \le cp, \qquad k = 0, 1, 2, 3. \tag{3.91}$$

With the estimates just obtained, we can apply global regularity estimate for the elastostatic system with Neumann boundary conditions (see [130], Theorem 6.6, p. 79) for $\boldsymbol{u}$ in $Q \setminus \overline{C}$, i.e.,

$$\|\boldsymbol{u}\|_{H^3(Q\setminus\overline{C})} \le c\Big(\|\boldsymbol{u}\|_{L^2(Q\setminus\overline{C})} + \|(\mathbb{C}\widehat{\nabla}\boldsymbol{u})\boldsymbol{n}\|_{H^{3/2}(\partial(Q\setminus\overline{C}))}\Big), \tag{3.92}$$

with $c = c(\lambda, \mu, \alpha, \beta, r_0, E_0, D_0)$. We estimate the two terms in the right-hand side of (3.92). Then, by (3.76)

$$\|\boldsymbol{u}\|_{L^2(Q\setminus\overline{C})} \le c\left\|\frac{\boldsymbol{u}}{\rho}\right\|_{L^2(\mathbb{R}^3_-\setminus\overline{C})} \le cp, \tag{3.93}$$

where $c = c(\lambda, \mu, \alpha, \beta, r_0, E_0, D_0)$. For the term $\|(\mathbb{C}\widehat{\nabla}\boldsymbol{u})\boldsymbol{n}\|_{H^{3/2}(\partial(Q\setminus\overline{C}))}$ we note that

$$(\mathbb{C}\widehat{\nabla}\boldsymbol{u})\boldsymbol{n} = \boldsymbol{0}, \quad \text{on } \partial Q \cap \{x_3 = 0\}, \qquad (\mathbb{C}\widehat{\nabla}\boldsymbol{u})\boldsymbol{n} = p\boldsymbol{n}, \quad \text{on } \partial C,$$

and since ∂C is of class C^3, see (3.78), it follows $(\mathbb{C}\widehat{\nabla}\boldsymbol{u})\boldsymbol{n} \in C^2(\partial C)$, hence, again

$$\|(\mathbb{C}\widehat{\nabla}\boldsymbol{u})\boldsymbol{n}\|_{H^{3/2}(\partial C)} \leq cp. \tag{3.94}$$

Analogously, from the estimate (3.91) and the regularity of the boundary of Q, we find

$$\|(\mathbb{C}\widehat{\nabla}\boldsymbol{u})\boldsymbol{n}\|_{H^{3/2}(\partial Q\setminus\partial C)} \leq cp. \tag{3.95}$$

Therefore, from (3.93)–(3.95), the estimate (3.92) gives

$$\|\boldsymbol{u}\|_{H^3(Q\setminus\overline{C})} \leq cp.$$

Finally, applying the general Sobolev embedding theorem, we get that $H^3(Q\setminus\overline{C}) \subset C^{1,1/2}(Q \setminus \overline{C})$, hence the assertion (3.88) follows. □

3.3.2 *Stability Estimates*

In this section, for any $\varrho > 0$, we denote by Ω_ϱ the set

$$\Omega_\varrho = \{\boldsymbol{x} \in \Omega \, : \, \mathrm{dist}(\boldsymbol{x}, \partial\Omega) > \varrho\}.$$

As explained at the beginning of this section, the proof of the stability estimates for the inverse problems is based on the use of different aspects of quantitative estimates of unique continuation. Specifically, the scheme to derive the stability estimates is to combine these two steps:

(a) the propagation of the smallness of the Cauchy data up to the boundary of the cavities, leading to an integral estimate of the solutions;
(b) an estimate of continuation from the interior.

The basic tool for both steps is the three-spheres inequality which is stated, for the Lamé system, in the following proposition

Lemma 3.3.2 (Three Spheres Inequality) *Let Ω be a bounded domain in $\mathbb{R}^3$. Let $\boldsymbol{w} \in H^1(\Omega)$ be a solution to the Lamé system. There exists ϑ^*, $0 < \vartheta^* \leq 1$, only depending on λ and μ such that for every $r_1, r_2, r_3, \overline{r}$, $0 < r_1 < r_2 < r_3 \leq \vartheta^*\overline{r}$, and for every $\boldsymbol{x} \in \Omega_{\overline{r}}$ we have*

$$\int_{B_{r_2}(\boldsymbol{x})} |\widehat{\nabla}\boldsymbol{w}|^2 \leq c \left(\int_{B_{r_1}(\boldsymbol{x})} |\widehat{\nabla}\boldsymbol{w}|^2 \right)^{\delta} \left(\int_{B_{r_3}(\boldsymbol{x})} |\widehat{\nabla}\boldsymbol{w}|^2 \right)^{1-\delta}, \tag{3.96}$$

where $c > 0$ and δ, $0 < \delta < 1$, only depend on λ, μ, $\frac{r_2}{r_3}$ and, increasingly, on $\frac{r_1}{r_3}$.

We do not provide the proof of this proposition which is now well-known in the literature of the inverse problem. For readers interested in this topic, we recommend, for instance, [4, 10].

Now, we state and prove an estimate of continuation from the interior which is also called *Lipschitz propagation of smallness.*

Proposition 3.3.3 (Lipschitz Propagation of Smallness) *Under the assumptions* (3.2), (3.48)–(3.50), *and* (3.78), *let $\boldsymbol{w}$ be the solution to* (3.83). *There exist $R \geq 3D_0$, $R = R(\lambda, \mu, r_0, E_0, D_0)$, and $s > 1$, $s = s(\lambda, \mu, E_0)$, such that for every $\varrho > 0$ and every $\overline{\boldsymbol{x}} \in (B_R^-(\mathbf{0}) \setminus \overline{C})_{s\varrho}$, we have*

$$\int_{B_\varrho(\overline{\boldsymbol{x}})} |\widehat{\nabla} \boldsymbol{w}|^2 \, d\boldsymbol{x} \geq \frac{c}{e^{a\varrho^{-b}}} \int_{B_R^-(\mathbf{0}) \setminus \overline{C}} |\widehat{\nabla} \boldsymbol{w}|^2 \, d\boldsymbol{x}, \tag{3.97}$$

where $a, b, c > 0$ depend on λ, μ, r_0, E_0, and D_0.

Proof For simplicity of notation, we denote $\Omega = B_R^-(\mathbf{0}) \setminus \overline{C}$, with $R \geq 3D_0$ to be chosen later. This proof is long and technical hence is divided in four steps. The first three steps come from [115] with some changes and simplifications in steps two and three.

The main idea of the proof is to cover the set Ω_{sp} with balls of variable radii (and chosen in order to be internally tangent to suitable cones contained in Ω) and then estimate the L^2-norm of $\widehat{\nabla} \boldsymbol{u}$ in these balls with $\int_{B_\varrho(\overline{\boldsymbol{x}})} |\widehat{\nabla} \boldsymbol{u}|^2 \, d\boldsymbol{x}$. This task is reached utilizing the three-spheres inequality of Lemma 3.3.2.

We note that if the half ball $B_1^-(\mathbf{0})$ has Lipschitz boundary with absolute constants r^*, E^*, then $B_R^-(\mathbf{0})$ has Lipschitz boundary with constants $r^* R$, E^*. Without loss of generality, we can assume $E_0 \geq E^*$ and $r_0 \leq r^* R$ (eventually worsening the regularity parameters of C, see (3.78)) so that the boundary of Ω is of Lipschitz class with constants r_0 and E_0. Again, for simplicity of notation, we can assume that $r_0 = 1$. The thesis of the theorem follows by a rescaling argument. Moreover, we assume that there exists $\varrho_0 = \varrho_0(\lambda, \mu, r_0, E_0, R)$, with $0 < \varrho_0 < 1$. Taking θ^* defined in Lemma 3.3.2, we define

$$\theta_0 = \arctan \frac{1}{E_0}, \tag{3.98}$$

$$s = \frac{5 + \theta^* \sin\theta_0 + \sqrt{(\theta^*)^2 \sin^2\theta_0 + 30\theta^* \sin\theta_0 + 25}}{2\theta^* \sin\theta_0} \tag{3.99}$$

$$\chi = \frac{s\theta^* \sin\theta_0}{5} \tag{3.100}$$

$$\vartheta_1 = \arcsin \frac{1}{s}. \tag{3.101}$$

We first note that the constants $s > 1$, $\chi > 1$, and $\vartheta_1 > 0$ depend on λ, μ, and E_0. In addition, given $z \in \mathbb{R}^3$, $\boldsymbol{\xi} \in \mathbb{R}^3$, with $|\boldsymbol{\xi}| = 1$, and $\vartheta > 0$, we define

$$K(z, \boldsymbol{\xi}, \vartheta) = \left\{ \boldsymbol{x} \in \mathbb{R}^3 \ : \ \frac{(\boldsymbol{x} - z) \cdot \boldsymbol{\xi}}{|\boldsymbol{x} - z|} > \cos \vartheta \right\}, \tag{3.102}$$

the open cone with vertex z, axis in the direction $\boldsymbol{\xi}$, and width 2ϑ.

Step 1 *For every ϱ, $0 < \varrho \leq \overline{\varrho} = \vartheta^*/16s$, and for every $\boldsymbol{x} \in \Omega$ satisfying $s\varrho < \operatorname{dist}(\boldsymbol{x}, \partial\Omega) \leq \vartheta^*/4$, there exists $\tilde{\boldsymbol{x}} \in \Omega$ such that*

(i) $B_{\frac{5\chi\varrho}{\vartheta^*}}(\boldsymbol{x}) \subset K(\tilde{\boldsymbol{x}}, \frac{\boldsymbol{x}-\tilde{\boldsymbol{x}}}{|\boldsymbol{x}-\tilde{\boldsymbol{x}}|}, \vartheta_0) \cap B_{\frac{\vartheta^*}{8}} \subset \Omega$.

(ii) The balls $B_\varrho(\boldsymbol{x})$ and $B_{\chi\varrho}(\boldsymbol{x}_2)$ are internally tangent to $K(\tilde{\boldsymbol{x}}, \frac{\boldsymbol{x}-\tilde{\boldsymbol{x}}}{|\boldsymbol{x}-\tilde{\boldsymbol{x}}|}, \vartheta_1)$, where $\boldsymbol{x}_2 = \boldsymbol{x} + (\chi + 1)\varrho \frac{\boldsymbol{x}-\tilde{\boldsymbol{x}}}{|\boldsymbol{x}-\tilde{\boldsymbol{x}}|}$.

Proof of Step 1 We denote by $d = \operatorname{dist}(\boldsymbol{x}, \partial\Omega)$ and by $\boldsymbol{w} \in \partial\Omega$ the point such that $|\boldsymbol{x} - \boldsymbol{w}| = d$. Without loss of generality, through a rigid transformation of coordinates, we can assume that $\boldsymbol{w} = \boldsymbol{0}$ and

$$\Omega \cap B_1(\boldsymbol{w}) = \{z \in B_1(\boldsymbol{0}) \ : \ z_3 > \psi(z')\},$$

where ψ is a Lipschitz function on $B_1(\boldsymbol{0})$, which satisfies

$$\psi(\boldsymbol{0}) = 0$$
$$\|\psi\|_{C^{0,1}(B_1(\boldsymbol{0}))} \leq E_0.$$

We define $\tilde{\boldsymbol{x}} = (\boldsymbol{x}', x_3 - s\varrho)$, hence we have that $|\boldsymbol{x} - \tilde{\boldsymbol{x}}| = s\varrho$, $|\tilde{\boldsymbol{x}} - \boldsymbol{w}| \leq d + s\varrho \leq 5/16\vartheta^*$. Utilizing the fact that ψ is Lipschitz, we find that $K(\tilde{\boldsymbol{x}}, \frac{\boldsymbol{x}-\tilde{\boldsymbol{x}}}{|\boldsymbol{x}-\tilde{\boldsymbol{x}}|}, \vartheta_0) \cap B_{\frac{\vartheta^*}{8}}(\tilde{\boldsymbol{x}}) \subset \Omega$. In addition, it holds

$$B_{\frac{5\chi\varrho}{\vartheta^*}} \subset K(\tilde{\boldsymbol{x}}, \frac{\boldsymbol{x} - \tilde{\boldsymbol{x}}}{|\boldsymbol{x} - \tilde{\boldsymbol{x}}|}, \vartheta_0) \cap B_{\frac{\vartheta^*}{8}}(\tilde{\boldsymbol{x}}),$$

if and only if

$$\frac{5\chi\varrho}{\vartheta^*} \leq s\varrho \sin\vartheta_0, \qquad \left(\frac{5\chi}{\vartheta^*} + s\right)\varrho \leq \frac{\vartheta^*}{8}.$$

Choosing χ as in (3.100) and using the fact that $\varrho \leq \overline{\varrho} = \vartheta^*/(16s)$, we get the assertion. Finally, we note that from the choice of ϑ_1, the ball $B_\varrho(\boldsymbol{x})$ is internally tangent to $K(\tilde{\boldsymbol{x}}, \frac{\boldsymbol{x}-\tilde{\boldsymbol{x}}}{|\boldsymbol{x}-\tilde{\boldsymbol{x}}|}, \vartheta_1)$, while $B_{\chi\varrho}(\boldsymbol{x}_2)$ is internally tangent to $K(\tilde{\boldsymbol{x}}, \frac{\boldsymbol{x}-\tilde{\boldsymbol{x}}}{|\boldsymbol{x}-\tilde{\boldsymbol{x}}|}, \vartheta_1)$ if and only if $\chi = (s + 1)/(s - 1)$. □

We set

$$\varrho_1 = \varrho, \qquad \varrho_k = \chi \varrho_{k-1} = \chi^{k-1}\varrho, \qquad k \geq 2,$$

$$\boldsymbol{x}_1 = \boldsymbol{x}, \qquad \boldsymbol{x}_k = \boldsymbol{x}_{k-1} + (\varrho_{k-1} + \varrho_k)\frac{\boldsymbol{x} - \tilde{\boldsymbol{x}}}{|\boldsymbol{x} - \tilde{\boldsymbol{x}}|}, \qquad k \geq 2.$$

We have that $B_{\varrho_k}(\boldsymbol{x}_k)$ is tangent from the interior to the cone $K(\tilde{\boldsymbol{x}}, \frac{\boldsymbol{x}-\tilde{\boldsymbol{x}}}{|\boldsymbol{x}-\tilde{\boldsymbol{x}}|}, \vartheta_1)$, while $B_{5\chi\varrho_k/\vartheta^*}(\boldsymbol{x}_k)$ to $K(\tilde{\boldsymbol{x}}, \frac{\boldsymbol{x}-\tilde{\boldsymbol{x}}}{|\boldsymbol{x}-\tilde{\boldsymbol{x}}|}, \vartheta_0)$. Moreover, it holds that $B_{5\varrho_k/\vartheta^*}(\boldsymbol{x}_k) \subset B_{\vartheta^*/8}(\tilde{\boldsymbol{x}})$ if and only if

$$k - 1 \leq \frac{\log\left[\frac{(\chi-1)\vartheta^*}{(5+\vartheta^*)(\chi-1)+2\vartheta^*}\left(\frac{\vartheta^*}{8\varrho} - s + 1 + \frac{2}{\chi-1}\right)\right]}{\ln \chi}.$$

To guarantee that $B_{5\varrho_k/\vartheta^*}(\boldsymbol{x}_k) \subset B_{\vartheta^*/8}(\tilde{\boldsymbol{x}})$ holds at least for $k = 1, 2$, we assume that $\varrho \leq \overline{\varrho}_1 = \frac{(\vartheta^*)^2}{8(\chi(5+\vartheta^*)+\vartheta^*(1+s))}$. We define

$$k(\varrho) = \left\lfloor \frac{\log\left[\frac{(\chi-1)\vartheta^*}{(5+\vartheta^*)(\chi-1)+2\vartheta^*}\left(\frac{\vartheta^*}{8\varrho} - s + 1 + \frac{2}{\chi-1}\right)\right]}{\ln \chi} \right\rfloor + 1,$$

where $\lfloor\cdot\rfloor$ represents the integer part of a real number. Therefore, we have that $B_{\frac{5\varrho_{k(\varrho)}}{\vartheta^*}}(\boldsymbol{x}_{k(\varrho)}) \subset B_{\frac{\vartheta^*}{8}}(\tilde{\boldsymbol{x}}) \cap \Omega$ and $B_{5\chi\varrho_i/\vartheta^*}(\boldsymbol{x}_i) \subset B_{\vartheta^*/8}(\tilde{\boldsymbol{x}}) \cap \Omega$, for $i = 1, \cdots, k(\varrho) - 1$. Since $\varrho \leq \frac{\vartheta^*}{16s}$, we get

$$k(\varrho) \geq \frac{\log \frac{\gamma}{\varrho}}{\log \chi}, \tag{3.103}$$

where $\gamma = \frac{(\vartheta^*)^2(\chi-1)}{16((5+\vartheta^*)(\chi-1)+2\vartheta^*)}$. Assuming also that $\varrho \leq \overline{\varrho}_2 = \frac{\vartheta^*(\chi-1)}{16}$, and since $\frac{\vartheta^*(\chi-1)}{(5+\vartheta^*)(\chi-1)+2\vartheta^*} \leq \frac{1}{5}$, we find that

$$k(\varrho) \leq \frac{\log \frac{\vartheta^*}{20\varrho}}{\log \chi} + 1. \tag{3.104}$$

Therefore, from (3.103) and (3.104) and for $\varrho \leq \min(\overline{\varrho}, \overline{\varrho}_1, \overline{\varrho}_2)$, we have

$$\frac{\gamma}{\chi} \leq \varrho_{k(\varrho)} = \chi^{k(\varrho)-1}\varrho \leq \frac{\vartheta^*}{20}. \tag{3.105}$$

We are ready now to state and prove the second step of the proof.

Step 2 *There exists $\tilde{\varrho} > 0$, which depends on λ, μ, and E_0, such that for every ϱ, with $0 < \varrho \le \tilde{\varrho}$, and for every $\boldsymbol{x} \in \Omega$, such that $s\varrho < \mathrm{dist}(\boldsymbol{x}, \partial\Omega) \le \frac{\vartheta^*}{4}$, it holds*

$$\frac{\int_{B_{\varrho_{k(\varrho)}}(\boldsymbol{x}_{k(\varrho)})} |\nabla \widehat{\boldsymbol{w}}|^2}{\int_\Omega |\nabla \widehat{\boldsymbol{w}}|^2} \le C \left(\frac{\int_{B_\varrho(\boldsymbol{x})} |\nabla \widehat{\boldsymbol{w}}|^2}{\int_\Omega |\nabla \widehat{\boldsymbol{w}}|^2} \right)^{\sigma_\chi^{k(\varrho)-1}}, \tag{3.106}$$

$$\frac{\int_{B_\varrho(\boldsymbol{x})} |\nabla \widehat{\boldsymbol{w}}|^2}{\int_\Omega |\nabla \widehat{\boldsymbol{w}}|^2} \le C \left(\frac{\int_{B_{\varrho_{k(\varrho)}}(\boldsymbol{x}_{k(\varrho)})} |\nabla \widehat{\boldsymbol{w}}|^2}{\int_\Omega |\nabla \widehat{\boldsymbol{w}}|^2} \right)^{\sigma^{k(\varrho)-1}}, \tag{3.107}$$

where $0 < \sigma < 1$ depends only on λ, μ, whereas $C > 1$ and $0 < \sigma_\chi < 1$ depend on λ, μ, and E_0.

Proof of Step 2 Let $\tilde{\varrho} = \min(\overline{\varrho}, \overline{\varrho}_1, \overline{\varrho}_2)$ and choose $\varrho \le \tilde{\varrho}$. We apply the three-spheres inequality, see Lemma 3.3.2, choosing balls centered at $\boldsymbol{x}_i$ and radii $r_1^{(i)} = \varrho_i$, $r_2^{(i)} = 3\chi\varrho_i$, $r_3^{(i)} = 4\chi\varrho_i$, for $i = 1, \cdots, k(\varrho) - 1$. Since $B_{r_1^{(i+1)}}(\boldsymbol{x}_{i+1}) \subset B_{r_2^{(i)}}(\boldsymbol{x}_i)$, for $i = 1, \cdots, k(\varrho) - 1$, we find that

$$\int_{B_{\varrho_{i+1}}(\boldsymbol{x}_{i+1})} |\nabla \widehat{\boldsymbol{w}}|^2 \le C \left(\int_{B_{\varrho_i}(\boldsymbol{x}_i)} |\nabla \widehat{\boldsymbol{w}}|^2 \right)^{\sigma_\chi} \left(\int_{B_{4\chi\varrho_i}(\boldsymbol{x}_i)} |\nabla \widehat{\boldsymbol{w}}|^2 \right)^{1-\sigma_\chi}, \tag{3.108}$$

where $C > 1$ and $0 < \sigma_\chi < 1$ depend on λ, μ, and E_0. Therefore (3.106) follows noticing that

$$\int_{B_{4\chi\varrho_i}(\boldsymbol{x}_i)} |\nabla \widehat{\boldsymbol{w}}|^2 \le \int_{B_{5\chi\varrho_i}(\boldsymbol{x}_i)} |\nabla \widehat{\boldsymbol{w}}|^2 \le \int_\Omega |\nabla \widehat{\boldsymbol{w}}|^2.$$

Indeed, from (3.108), we have

$$\frac{\int_{B_{\varrho_{i+1}}(\boldsymbol{x}_{i+1})} |\nabla \widehat{\boldsymbol{w}}|^2}{\int_\Omega |\nabla \widehat{\boldsymbol{w}}|^2} \le \left(\frac{\int_{B_{\varrho_i}(\boldsymbol{x}_i)} |\nabla \widehat{\boldsymbol{w}}|^2}{\int_\Omega |\nabla \widehat{\boldsymbol{w}}|^2} \right)^{\sigma_\chi},$$

hence the assertion follows by iterating, for $i = 1, \cdots, k(\varrho) - 1$, the last inequality. Equation (3.107) follows similarly applying the three-spheres inequality to the balls $B_{\varrho_i}(\boldsymbol{x}_i)$, $B_{3\varrho_i}(\boldsymbol{x}_i)$, $B_{4\varrho_i}(\boldsymbol{x}_i)$ for $i = 1, \cdots, k(\varrho) - 1$, noticing that $B_{\varrho_i}(\boldsymbol{x}_{i-1}) \subset B_{3\varrho_i}(\boldsymbol{x}_i)$ and repeating the argument above. □

Step 3 *There exists ϱ_0, which depends on λ, μ, E_0, and R, such that for every $\varrho \le \varrho_0$ and $\overline{\boldsymbol{x}} \in \Omega_{s\varrho}$ it holds*

$$\int_{B_\varrho(\overline{x})} |\widehat{\nabla w}|^2\,dx \geq c_1 \int_\Omega |\widehat{\nabla w}|^2\,dx \left(\frac{c\varrho^3 \int_{\Omega_{(s+1)\varrho}} |\widehat{\nabla w}|^2\,dx}{\int_\Omega |\widehat{\nabla w}|^2\,dx} \right)^{\sigma_\chi^{-A_1-B_1\log(1/\varrho)}}, \tag{3.109}$$

where $c_1 > 0$ *only depend on* λ, μ; $\sigma_\chi \in (0,1), s > 1$ *depend on* λ, μ, E_0 *and* $c, A_1, B_1 > 0$ *depend on* $\lambda, \mu, E_0, r_0, R$.

Proof of Step 3 We prove the thesis of Step 3. For every point $\overline{x} \in \Omega_{s\varrho}$ satisfying $\text{dist}(\overline{x}, \partial\Omega) \leq \vartheta^*/4$. We do not provide the proof for the case $\text{dist}(\overline{x}, \partial\Omega) > \vartheta^*/4$ since it can be proved similarly to the other case and with less effort. The proof of this second case can be found, for instance, in [115], see Step 4. at page 469.
Case $\overline{x} \in \Omega_{s\varrho}$ *satisfying* $\text{dist}(\overline{x}, \partial\Omega) \leq \vartheta^*/4$. Let ϱ be such that $\varrho \leq \tilde{\varrho}$. From (3.105), we have that $\frac{5\varrho_{k(\varrho)}}{\vartheta^*} < \frac{1}{3}$, hence it follows that $\Omega_{\frac{5\varrho_{k(\varrho)}}{\vartheta^*}}$ is connected. We choose $y \in \Omega$ such that $s\varrho < \text{dist}(y, \partial\Omega) \leq \frac{\vartheta^*}{4}$ and let $\gamma \in \Omega_{\frac{5\varrho_{k(\varrho)}}{\vartheta^*}}$ be an arc joining $\overline{x}_{k(\varrho)}$ to $y_{k(\varrho)}$. Then, we define the following process: let x_i, for $i = 1, \cdots, L$, be such that $x_1 = \overline{x}$ and $x_{i+1} = \gamma(t_i)$, where $t_i = \max\{t : |\gamma(t) - x_i| = 2\varrho_{k(\varrho)}\}$ if $|x_i - y_{k(\varrho)}| > 2\varrho k(\varrho)$; otherwise let $i = L$ and stop the process. By construction, the balls $B_{\varrho_{k(\varrho)}}$ are pairwise disjoint and $|x_{i+1} - x_i| = 2\varrho_{k(\varrho)}$, for $i = 1, \cdots, L-1$, and $|x_L - y_{k(\varrho)}| \leq 2\varrho_{k(\varrho)}$. Therefore

$$L \leq \frac{3R}{4\pi\varrho_{k(\varrho)}^3}. \tag{3.110}$$

As in the previous step, applying iteratively the three-spheres inequality to the balls centered in x_i with radii $r_1 = \varrho_{k(\varrho)}$, $r_2 = 3\varrho_{k(\varrho)}$, and $r_3 = 4\varrho_{k(\varrho)}$, we find

$$\frac{\int_{B_{\varrho_{k(\varrho)}}(y_{k(\varrho)})} |\widehat{\nabla w}|^2}{\int_\Omega |\widehat{\nabla w}|^2} \leq C \left(\frac{\int_{B_{\varrho_{k(\varrho)}}(\overline{x}_{k(\varrho)})} |\widehat{\nabla w}|^2}{\int_\Omega |\widehat{\nabla w}|^2} \right)^{\sigma^L}, \tag{3.111}$$

where $C > 1$ depends on λ and μ. At the same time, choosing $x = \overline{x}$ in (3.106) and $x = y$ in (3.107), we get

$$\frac{\int_{B_{\varrho_{k(\varrho)}}(\overline{x}_{k(\varrho)})} |\widehat{\nabla w}|^2}{\int_\Omega |\widehat{\nabla w}|^2} \leq C \left(\frac{\int_{B_\varrho(\overline{x})} |\widehat{\nabla w}|^2}{\int_\Omega |\widehat{\nabla w}|^2} \right)^{\sigma_\chi^{k(\varrho)-1}}, \tag{3.112}$$

$$\frac{\int_{B_\varrho(y)} |\widehat{\nabla w}|^2}{\int_\Omega |\widehat{\nabla w}|^2} \leq C \left(\frac{\int_{B_{\varrho_{k(\varrho)}}(y_{k(\varrho)})} |\widehat{\nabla w}|^2}{\int_\Omega |\widehat{\nabla w}|^2} \right)^{\sigma^{k(\varrho)-1}}, \tag{3.113}$$

where $C > 1$ depends on λ, μ, and E_0. From (3.111)–(3.113) we find

$$\frac{\int_{B_\varrho(y)} |\widehat{\nabla} \boldsymbol{w}|^2}{\int_\Omega |\widehat{\nabla} \boldsymbol{w}|^2} \leq C \left(\frac{\int_{B_\varrho(\overline{x})} |\widehat{\nabla} \boldsymbol{w}|^2}{\int_\Omega |\widehat{\nabla} \boldsymbol{w}|^2} \right)^{\sigma_\chi^{k(\varrho)-1} \sigma^{k(\varrho)+L-1}}, \tag{3.114}$$

where $C > 1$ depends on λ, μ, and E_0.

Next, we consider the case where $\boldsymbol{y} \in \Omega$ is such that $\mathrm{dist}(\boldsymbol{y}, \partial\Omega) > \frac{\vartheta^*}{4}$. Let $\tilde{r} = \vartheta^* \varrho_{k(\varrho)}$, then, by (3.105), we get

$$\mathrm{dist}(\overline{\boldsymbol{x}}_{k(\varrho)}, \partial\Omega) \geq \frac{5}{\vartheta^*} \varrho_{k(\varrho)} > \frac{5}{\vartheta^*} \tilde{r}$$

$$\mathrm{dist}(\boldsymbol{y}, \partial\Omega) > \frac{\vartheta^*}{4} \geq 5\varrho_{k(\varrho)} = \frac{5}{\vartheta^*} \tilde{r}.$$

Since $\tilde{r} < \varrho_{k(\varrho)}$, the domain $\Omega_{\varrho_{k(\varrho)}}$ is connected. We can now apply a similar argument as above, taking an arc joining $\overline{\boldsymbol{x}}_{k(\varrho)}$ and $\boldsymbol{y}$ and applying three-spheres inequality over a chain of $\widetilde{L}$ balls centered at $\boldsymbol{x}_i \in \Omega_{\frac{5}{\vartheta^*}\tilde{r}}$ and radii $\tilde{r}$, $3\tilde{r}$, $4\tilde{r}$, where

$$\widetilde{L} \leq \frac{3R}{4\pi \tilde{r}^3}. \tag{3.115}$$

Let ϱ be such that $\varrho \leq \overline{\varrho}_3 = \frac{\gamma \vartheta^*}{\chi}$, such that $\varrho \leq \tilde{r}$ hence

$$\int_{B_{4\tilde{r}}(\boldsymbol{x}_i)} |\widehat{\nabla} \boldsymbol{w}|^2 \leq \int_{B_{5\tilde{r}}(\boldsymbol{x}_i)} |\widehat{\nabla} \boldsymbol{w}|^2 \leq \int_\Omega |\widehat{\nabla} \boldsymbol{w}|^2,$$

therefore from the three-spheres inequality, we find

$$\frac{\int_{B_{\tilde{r}}(y)} |\widehat{\nabla} \boldsymbol{w}|^2}{\int_\Omega |\widehat{\nabla} \boldsymbol{w}|^2} \leq C \left(\frac{\int_{B_{\tilde{r}}(\overline{\boldsymbol{x}}_{k(\varrho)})} |\widehat{\nabla} \boldsymbol{w}|^2}{\int_\Omega |\widehat{\nabla} \boldsymbol{w}|^2} \right)^{\sigma^{\widetilde{L}}}. \tag{3.116}$$

Putting together (3.112) and (3.116) and recalling that $\varrho \leq \tilde{r} < \varrho_{k(\varrho)}$, we find

$$\frac{\int_{B_\varrho(y)} |\widehat{\nabla} \boldsymbol{w}|^2}{\int_\Omega |\widehat{\nabla} \boldsymbol{w}|^2} \leq C \left(\frac{\int_{B_\varrho(\overline{x})} |\widehat{\nabla} \boldsymbol{w}|^2}{\int_\Omega |\widehat{\nabla} \boldsymbol{w}|^2} \right)^{\sigma_\chi^{k(\varrho)-1} \sigma^{\widetilde{L}}}, \tag{3.117}$$

where $C > 1$ depends on λ, μ, and E_0. Therefore, utilizing (3.117), (3.114), (3.104), (3.110), (3.115), and recalling that $\sigma_\chi < \sigma$, we get, for every $\boldsymbol{y} \in \Omega_{s\varrho}$, that

$$\frac{\int_{B_\varrho(y)}|\widehat{\nabla}\boldsymbol{w}|^2}{\int_\Omega|\widehat{\nabla}\boldsymbol{w}|^2} \le C\left(\frac{\int_{B_\varrho(\overline{x})}|\widehat{\nabla}\boldsymbol{w}|^2}{\int_\Omega|\widehat{\nabla}\boldsymbol{w}|^2}\right)^{\sigma_\chi^{A_1+B_1\log(1/\varrho)}}, \tag{3.118}$$

where $C > 1$ depends on λ, μ and E_0, while A_1, B_1 depend on λ, μ, E_0 and R.

We now cover $\Omega_{(s+1)\varrho}$ with internally nonoverlapping closed cubes of side $l = \frac{2\varrho}{\sqrt{3}}$. Any cube is contained in a ball centered at a point of $\Omega_{s\varrho}$ and radius ϱ. Moreover the number of cubes has an upper bound given by

$$N \le \frac{R3^{(3/2)}}{8\varrho^3}. \tag{3.119}$$

Therefore, from (3.118) and (3.119), we find

$$\int\limits_{B_\varrho(\overline{x})}|\widehat{\nabla}\boldsymbol{w}|^2\,d\boldsymbol{x} \ge c_1\int\limits_\Omega|\widehat{\nabla}\boldsymbol{w}|^2\,d\boldsymbol{x}\left(\frac{c\varrho^3\int\limits_{\Omega_{(s+1)\varrho}}|\widehat{\nabla}\boldsymbol{w}|^2\,d\boldsymbol{x}}{\int\limits_\Omega|\widehat{\nabla}\boldsymbol{w}|^2\,d\boldsymbol{x}}\right)^{\sigma_\chi^{-A_1-B_1\log(1/\varrho)}}, \tag{3.120}$$

where $c_1 > 0$ only depend on λ, μ; $\sigma_\chi \in (0, 1)$, $s > 1$ depend on λ, μ, E_0 and c, A_1, $B_1 > 0$ depend on λ, μ, E_0, R. □

It remains to prove, to get the assertion of the theorem, that

Step 4 For every ϱ such that $0 < \varrho \le \varrho_0$ and for every $\overline{x} \in \Omega_{s\varrho}$, we have that

$$\frac{\int\limits_{\Omega_{(s+1)\varrho}}|\widehat{\nabla}\boldsymbol{w}|^2\,d\boldsymbol{x}}{\int\limits_\Omega|\widehat{\nabla}\boldsymbol{w}|^2\,d\boldsymbol{x}} \ge \frac{1}{2}.$$

Proof of Step 4 It is straightforward to observe that

$$\frac{\int\limits_{\Omega_{(s+1)\varrho}}|\widehat{\nabla}\boldsymbol{w}|^2\,d\boldsymbol{x}}{\int\limits_\Omega|\widehat{\nabla}\boldsymbol{w}|^2\,d\boldsymbol{x}} = 1 - \frac{\int\limits_{\Omega\setminus\Omega_{(s+1)\varrho}}|\widehat{\nabla}\boldsymbol{w}|^2\,d\boldsymbol{x}}{\int\limits_\Omega|\widehat{\nabla}\boldsymbol{w}|^2\,d\boldsymbol{x}} := 1 - \frac{\mathcal{I}_1}{\mathcal{I}_2}. \tag{3.121}$$

Therefore, we estimate the ratio $\mathcal{I}_1/\mathcal{I}_2$. To get an estimate of $\mathcal{I}_2$, we make use of the fact that $\boldsymbol{w} = \boldsymbol{u} - \overline{\boldsymbol{u}}$, see (3.82), the integral representation formula (3.33) of the

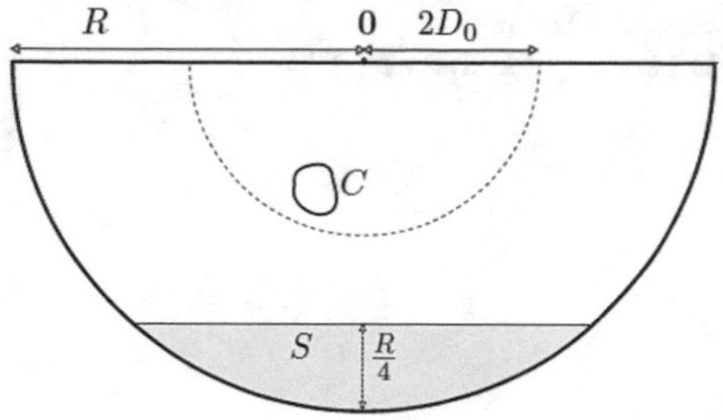

Fig. 3.1 The region S. Courtesy of [37]

function $\boldsymbol{u}$ and the explicit expression of $\overline{\boldsymbol{u}}$ in (3.81). Moreover, we fix the set

$$S = \left\{ \boldsymbol{x} \in B_R^-(\mathbf{0}) \ : \ x_3 \leq -\frac{3}{4} R \right\}. \tag{3.122}$$

see Fig. 3.1, which has measure $|S| = (7/128)\pi R^3$.

For any $\boldsymbol{y} \in S$ and $\boldsymbol{x} \in \partial C$, utilizing (3.33) and Theorem 3.1.7, it follows that

$$|\widehat{\nabla} \boldsymbol{u}(\boldsymbol{y})| \leq |\nabla \boldsymbol{u}(\boldsymbol{y})| \leq \frac{cp}{R^2}, \quad \forall \boldsymbol{y} \in S,$$

where $c = c(\lambda, \mu, r_0, E_0, D_0)$, hence

$$|\widehat{\nabla} \boldsymbol{w}(\boldsymbol{y})| \geq |\widehat{\nabla} \overline{\boldsymbol{u}}(\boldsymbol{y})| - |\widehat{\nabla} \boldsymbol{u}(\boldsymbol{y})| \geq \frac{p}{3\lambda + 2\mu} - \frac{cp}{R^2} \geq \frac{p}{2(3\lambda + 2\mu)},$$

where the last inequality holds choosing $R = \max\{3D_0, (2c^{-1}(3\lambda + 2\mu))^{1/2}\}$. Therefore, we find that

$$\mathcal{I}_2 = \int_\Omega |\widehat{\nabla} \boldsymbol{w}|^2 \, d\boldsymbol{x} \geq \int_S |\widehat{\nabla} \boldsymbol{w}|^2 \, d\boldsymbol{x} \geq c p^2 R^3, \tag{3.123}$$

where $c = c(\lambda, \mu)$.

Next, utilizing the regularity result of the Proposition 3.3.1, we get an estimate of the integral $\mathcal{I}_1$. First, we consider the following decomposition of the domain

$$\Omega \setminus \Omega_{(s+1)p} = F_1 \cup F_2,$$

where

$$F_1 = \{\boldsymbol{x} \in \Omega \ : \ d(\boldsymbol{x}, C)) \leq (s+1)\varrho\}, \qquad F_2 = \{\boldsymbol{x} \in \Omega \ : \ d(\boldsymbol{x}, \partial B_R^-(\mathbf{0})) \leq (s+1)\varrho\},$$

see Fig. 3.2.

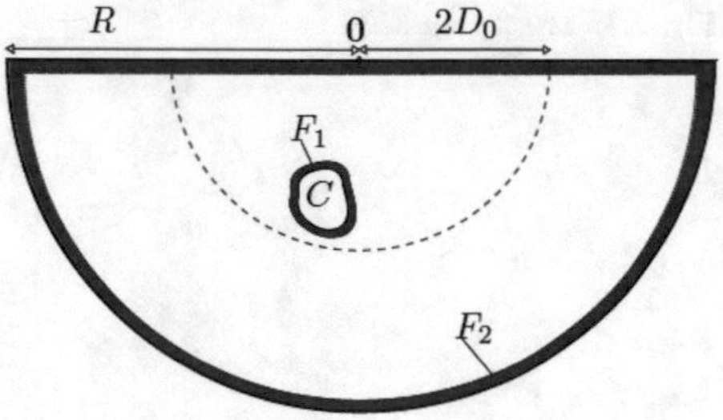

Fig. 3.2 The region $\Omega \setminus \Omega_{(s+1)\varrho}$. Courtesy of [37]

From (3.78), we notice that

$$|F_1 \cup F_2| \leq c(r_0, E_0, D_0)\varrho R^2. \tag{3.124}$$

We take $\alpha = R/D_0$ in (3.87), where $R = \max\{3D_0, (2c^{-1}(3\lambda + 2\mu))^{1/2}\}$, and $\beta = 2\alpha$, hence

$$\overline{B_R^-(\mathbf{0})} \subset\subset \overline{Q} \subset\subset \overline{B_{2R}^-(\mathbf{0})},$$

From the regularity results (3.88), it follows that

$$\|\widehat{\nabla}\boldsymbol{w}\|_{L^\infty(\Omega\setminus\Omega_{(s+1)\varrho})} \leq \left(\|\widehat{\nabla}\boldsymbol{u}\|_{L^\infty(\Omega\setminus\Omega_{(s+1)\varrho})} + \|\widehat{\nabla}\overline{\boldsymbol{u}}\|_{L^\infty(\Omega\setminus\Omega_{(s+1)\varrho})}\right) \leq cp,$$

where $c = c(\lambda, \mu, E_0, D_0)$. Therefore, from (3.124), we find

$$\mathcal{I}_1 = \int\limits_{\Omega\setminus\Omega_{(s+1)\varrho}} |\widehat{\nabla}\boldsymbol{w}|^2\, d\boldsymbol{x} \leq cp^2\, |\Omega \setminus \Omega_{(s+1)\varrho}| \leq cp^2\varrho R^2. \tag{3.125}$$

Collecting (3.123) and (3.125), we find that there exists $\varrho_0 = \varrho_0(\lambda, \mu, E_0, D_0) > 0$, such that for any $\varrho \leq \varrho_0$, we have

$$\frac{\mathcal{I}_1}{\mathcal{I}_2} = \frac{\int\limits_{\Omega\setminus\Omega_{(s+1)\varrho}} |\widehat{\nabla}\boldsymbol{w}|^2\, d\boldsymbol{x}}{\int\limits_{\Omega} |\widehat{\nabla}\boldsymbol{w}|^2\, d\boldsymbol{x}} \leq \frac{1}{2}.$$

□

Finally, utilizing this last result in (3.121) and then going back to (3.120), we find the assertion of the theorem, indeed,

$$\int\limits_{B_\varrho(\overline{\boldsymbol{x}})} |\widehat{\nabla}\boldsymbol{w}|^2\, d\boldsymbol{x} \geq \left(c\varrho^3\right)^{\sigma^{-A_1 - B_1 \log(1/\varrho)}} \int\limits_{\Omega} |\widehat{\nabla}\boldsymbol{w}|^2\, d\boldsymbol{x},$$

where c, A_1, B_1 depend on λ, μ, E_0, D_0, for all $\varrho \leq \varrho_0$. To conclude, we take $\varrho \leq c$, hence, for every $\varrho \leq \min(c, \varrho_0)$ and noticing that $\log \varrho \geq -1/\varrho$ for $0 < \varrho < 1$, we get the assertion choosing

$$a = 6e^{A_1|\log\sigma|} \qquad \text{and} \qquad b = B_1|\log\sigma| + 1.$$

We skip the proof of the following two propositions. In fact they are Propositions 3.5 and 3.6 in [115] in which we take as Neumann boundary condition pe_3.

Proposition 3.3.4 (Stability Estimates of Continuation from Cauchy Data) *Given the assumption* (3.2), *we consider two domains C_1 and C_2 satisfying* (3.48)–(3.50) *and* (3.78). *Then, taking $\boldsymbol{w}_i$, for $i = 1, 2$, solutions to* (3.83) *with $C = C_i$, we have, for $\varepsilon < e^{-1}p$, that*

$$\begin{aligned} \int_{C_2\setminus\overline{C}_1} |\widehat{\nabla}\boldsymbol{w}_1|^2\, d\boldsymbol{x} \leq cp^2\left(\log\left|\log\frac{\varepsilon}{p}\right|\right)^{-1/6}, \\ \int_{C_1\setminus\overline{C}_2} |\widehat{\nabla}\boldsymbol{w}_2|^2\, d\boldsymbol{x} \leq cp^2\left(\log\left|\log\frac{\varepsilon}{p}\right|\right)^{-1/6}, \end{aligned} \tag{3.126}$$

where the constant $c = c(\lambda, \mu, r_0, E_0, D_0, s_0)$.

The stability estimates in the previous theorem, i.e., (3.126) can be improved when ∂G is of Lipschitz class, where G is defined by (3.79), as stated in the proposition below.

Proposition 3.3.5 (Improved Stability Estimates of Continuation from Cauchy Data) *Under the assumption* (3.2) *let C_1 and C_2 be two domains satisfying* (3.48)–(3.50) *and* (3.78). *In addition, let us assume that there exist $L > 0$ and $\widetilde{r}_0$, with $0 < \widetilde{r}_0 \leq r_0$, such that ∂G is of Lipschitz class with constants $\widetilde{r}_0$, L. Then, we have*

$$\begin{aligned} \int_{C_2\setminus\overline{C}_1} |\widehat{\nabla}\boldsymbol{w}_1|^2\, d\boldsymbol{x} \leq cp^2\left|\log\frac{\varepsilon}{p}\right|^{-\gamma}, \\ \int_{C_1\setminus\overline{C}_2} |\widehat{\nabla}\boldsymbol{w}_2|^2\, d\boldsymbol{x} \leq cp^2\left|\log\frac{\varepsilon}{p}\right|^{-\gamma}, \end{aligned} \tag{3.127}$$

where $c, \gamma > 0$ depend on $\lambda, \mu, r_0, E_0, D_0, s_0, L, \widetilde{r}_0$.

Finally, we state and prove the stability theorem related to our inverse problem.

Theorem 3.3.6 (Stability Estimate) *Under the assumption* (3.2) *let C_1 and C_2 be two domains satisfying* (3.48)–(3.50) *and* (3.78). *Moreover, let $\boldsymbol{u}_i$, for $i = 1, 2$, be the solution to* (3.46) *with $C = C_i$. If, given $\varepsilon > 0$, we have*

$$\|\boldsymbol{u}_1 - \boldsymbol{u}_2\|_{L^2(B'_{s_0}(\mathbf{0}))} \leq \varepsilon, \tag{3.128}$$

then it holds

$$d_{\mathcal{H}}(\partial C_1, \partial C_2) \leq c\left(\log\left|\log\frac{\varepsilon}{p}\right|\right)^{-\eta}, \tag{3.129}$$

for every $\varepsilon < e^{-1}p$, *where the constants* c *and* η, *with* $0 < \eta \leq 1$, *depend on* $\lambda, \mu, r_0, E_0, D_0$ *and* s_0.

Proof Thanks to (3.85), we can prove the assertion working with the function $\boldsymbol{w}_i$, for $i = 1, 2$. In this way, we repeat the same procedure contained in [116]. In the sequel we simply denote with $d_{\mathcal{H}}$ the Hausdorff distance between the two sets ∂C_1 and ∂C_2, i.e.,

$$d_{\mathcal{H}}(\partial C_1, \partial C_2) = \max\Big\{\sup_{\boldsymbol{x}\in\partial C_1}\inf_{\boldsymbol{y}\in\partial C_2}|\boldsymbol{x}-\boldsymbol{y}|, \sup_{\boldsymbol{y}\in\partial C_2}\inf_{\boldsymbol{x}\in\partial C_1}|\boldsymbol{x}-\boldsymbol{y}|\Big\}.$$

We need to prove that if $\eta > 0$ is such that

$$\int_{C_2\setminus\overline{C}_1} |\widehat{\nabla}\boldsymbol{w}_1|^2\, d\boldsymbol{x} \leq \eta, \qquad \int_{C_1\setminus\overline{C}_2} |\widehat{\nabla}\boldsymbol{w}_2|^2\, d\boldsymbol{x} \leq \eta, \tag{3.130}$$

then it holds

$$d_{\mathcal{H}} \leq c\left(\log\frac{cp^2}{\eta}\right)^{-1/b}, \tag{3.131}$$

where b, c depend on $\lambda, \mu, r_0, E_0, D_0$.

With no loss of generality, we can always assume that there exists $\boldsymbol{x}_0 \in \partial C_1$ such that

$$\mathrm{dist}(\boldsymbol{x}_0, \partial C_2) = d_{\mathcal{H}}.$$

In this setting, we have to distinguish two cases:

(i) $B_{d_{\mathcal{H}}}(\boldsymbol{x}_0) \subset C_2$;
(ii) $B_{d_{\mathcal{H}}}(\boldsymbol{x}_0) \cap C_2 = \emptyset$.

Let us consider case (i). By the regularity assumption made on ∂C_1, see (3.78), there exists $\boldsymbol{x}_1 \in C_2 \setminus \overline{C}_1$ such that

$$B_{td}(\boldsymbol{x}_1) \subset (C_2 \setminus \overline{C}_1), \text{ with } t = \frac{1}{1+\sqrt{1+E_0^2}}.$$

By the first inequality in (3.130), taking $\varrho = td_{\mathcal{H}}/s$ in Proposition 3.3.3, we have

$$\eta \geq \int_{C_2\setminus\overline{C}_1} |\widehat{\nabla}\boldsymbol{w}_1|^2\, d\boldsymbol{x} \geq \int_{B_\varrho(\boldsymbol{x}_1)} |\widehat{\nabla}\boldsymbol{w}_1|^2\, d\boldsymbol{x} \geq \frac{c}{e^{a\varrho^{-b}}} \int_{B_{\overline{R}}\setminus\overline{C}_1} |\widehat{\nabla}\boldsymbol{w}_1|^2\, d\boldsymbol{x}, \qquad (3.132)$$

where we recall that $R = R(\lambda, \mu, r_0, E_0, D_0)$. By (3.123), we find that

$$\int_{B_{\overline{R}}\setminus\overline{C}_1} |\widehat{\nabla}\boldsymbol{w}_1|^2\, d\boldsymbol{x} \geq cp^2,$$

so that, going back to (3.132), we have

$$\eta \geq \frac{cp^2}{e^{a\varrho^{-b}}} = \frac{cp^2}{e^{a(td_{\mathcal{H}}/s)^{-b}}}. \qquad (3.133)$$

From this inequality it is straightforward to find (3.131).

Case (ii) can be proved in a similar way by substituting $\boldsymbol{w}_1$ with $\boldsymbol{w}_2$ in the previous calculations and employing the second inequality in (3.130).

Now, applying (3.126), that is taking

$$\eta = cp^2\left(\log\left|\log\frac{\varepsilon}{p}\right|\right)^{-1/6},$$

we obtain from (3.131) that

$$d_{\mathcal{H}} \leq c\left(\log\log\left|\log\frac{\varepsilon}{p}\right|\right)^{-1/b}, \qquad (3.134)$$

where we require $\varepsilon < e^{-e}p$ to have a positive quantity in right side of the previous inequality; the positive constants b, c depend on $\lambda, \mu, r_0, E_0, s_0$ and D_0.

Next, to improve the modulus of continuity of this estimate we recall a geometrical result, first introduced and proved in [7], ensuring that there exists $d_0 > 0$, $d_0 = d_0(r_0, E_0)$ such that if $d_{\mathcal{H}}(\partial C_1, \partial C_2) \leq d_0$, then the boundary of G is of Lipschitz class with constants $\widetilde{r}_0$, L, only depending on r_0 and E_0. By (3.134), there exists $\varepsilon_0 > 0$ only depending on $\lambda, \mu, r_0, E_0, s_0$, and D_0 such that if $\varepsilon \leq \varepsilon_0$ then $d_{\mathcal{H}} \leq d_0$. In this way G satisfies the hypotheses of Proposition 3.3.5 hence the assertion follows. □

In the next section we find the asymptotic expansion of the solution of the direct problem (3.31) when the cavity is small compared to the distance from the boundary of the half-space.

3.4 Asymptotic Expansion

As explained in the introduction, one way to restore the well-posedness of the inverse problem, is to add some a priori information. In particular, in this section, we consider the hypothesis that the cavity C of the elastic problem

$$\begin{cases} \operatorname{div}(\mathbb{C}\widehat{\nabla}\boldsymbol{u}) = \boldsymbol{0} & \text{in } \mathbb{R}^3_- \setminus C \\ \dfrac{\partial \boldsymbol{u}}{\partial \boldsymbol{\nu}} = p\,\boldsymbol{n} & \text{on } \partial C \\ \dfrac{\partial \boldsymbol{u}}{\partial \boldsymbol{\nu}} = \boldsymbol{0} & \text{on } \mathbb{R}^2 \\ \boldsymbol{u} = o(\mathbf{1}), \quad \nabla \boldsymbol{u} = o(|\boldsymbol{x}|^{-1}) & |\boldsymbol{x}| \to \infty, \end{cases} \tag{3.135}$$

is of the form

$$C = d_0 z + r_0 \Omega,$$

where $d_0, r_0 > 0$, $d_0 z$ is the center of the cavity and Ω is a bounded Lipschitz domain containing the origin. We assume that $d_0 >> r_0$ which means that the cavity is small compared to the distance from the boundary of the half-space. As proved by Friedman and Vogelius in [85], with this assumption one can retrieve a Lipschitz stability estimate for the inverse problem of detecting the cavity.

We rescale the variables $(\boldsymbol{x}, \boldsymbol{u}) \to (\boldsymbol{x}/d_0, \boldsymbol{u}/r0)$ hence defining by $\varepsilon := r_0/d_0$ and $C_\varepsilon := z + \varepsilon\Omega$ we denote the new variables by $\boldsymbol{u}_\varepsilon$ and $\boldsymbol{x}$. We find that (3.135) can be rewritten as

$$\begin{cases} \operatorname{div}(\mathbb{C}\widehat{\nabla}\boldsymbol{u}_\varepsilon) = \boldsymbol{0} & \text{in } \mathbb{R}^3_- \setminus \overline{C}_\varepsilon \\ \dfrac{\partial \boldsymbol{u}_\varepsilon}{\partial \boldsymbol{\nu}} = p\,\boldsymbol{n} & \text{on } \partial C_\varepsilon \\ \dfrac{\partial \boldsymbol{u}_\varepsilon}{\partial \boldsymbol{\nu}} = \boldsymbol{0} & \text{on } \mathbb{R}^2 \\ \boldsymbol{u}_{\boldsymbol{\varepsilon}} = o(\mathbf{1}), \quad \nabla \boldsymbol{u}_{\boldsymbol{\varepsilon}} = o(|\boldsymbol{x}|^{-1}) & |\boldsymbol{x}| \to \infty, \end{cases} \tag{3.136}$$

where, now, the quantity p is a rescaled pressure of the original pressure p and ε. Using the results of Sect. 3.2.1, i.e. the representation formula (3.33), we represent the solution of the boundary value problem (3.136) as

$$\begin{aligned} u^k_\varepsilon(\boldsymbol{y}) &= p \int\limits_{\partial C_\varepsilon} \boldsymbol{N}^{(k)}(\boldsymbol{x}, \boldsymbol{y}) \cdot \boldsymbol{n}(\boldsymbol{x})\, d\sigma(\boldsymbol{x}) - \int\limits_{\partial C_\varepsilon} \frac{\partial \boldsymbol{N}^{(k)}}{\partial \boldsymbol{\nu}}(\boldsymbol{x}, \boldsymbol{y}) \cdot \boldsymbol{f}(\boldsymbol{x})\, d\sigma(\boldsymbol{x}) \\ &:= I_1^{(k)}(\boldsymbol{y}) + I_2^{(k)}(\boldsymbol{y}), \quad \boldsymbol{y} \in \mathbb{R}^2 \end{aligned} \tag{3.137}$$

for $k = 1, 2, 3$, where u_ε^k indicates the k-th component of the displacement vector and $\boldsymbol{f}$ is the solution of (3.34), that is

$$\left(\tfrac{1}{2}\mathbf{I} + \mathbf{K}_\varepsilon + \mathbf{D}_\varepsilon^R\right) \boldsymbol{f}_\varepsilon(\boldsymbol{x}) = p\left(\mathbf{S}_\varepsilon^\Gamma(\boldsymbol{n})(\boldsymbol{x}) + \mathbf{S}_\varepsilon^R(\boldsymbol{n})(\boldsymbol{x})\right), \qquad \boldsymbol{x} \in \partial C_\varepsilon. \tag{3.138}$$

Obviously the dependence from ε in all layer potentials needs in the sequel to distinguish them from the layer potential defined over a domain independent from ε.

For any fixed value of $\varepsilon > 0$ and given $\boldsymbol{h} : \partial C_\varepsilon \to \mathbb{R}^3$, as did in the Chap. 2, Sect. 2.5, we introduce the function $\boldsymbol{h}^\sharp : \partial\Omega \to \mathbb{R}^3$ which is defined by

$$\boldsymbol{h}^\sharp(\boldsymbol{\zeta}) := \boldsymbol{h}(z + \varepsilon\boldsymbol{\zeta}), \qquad \boldsymbol{\zeta} \in \partial\Omega.$$

For the asymptotic result, a key role is assumed by the following auxiliary functions $\boldsymbol{\theta}^{qr}$, for $q, r = 1, 2, 3$, solutions to

$$\operatorname{div}(\mathbb{C}\widehat{\nabla}\boldsymbol{\theta}^{qr}) = 0 \quad \text{in } \mathbb{R}^3 \setminus \Omega, \qquad \frac{\partial \boldsymbol{\theta}^{qr}}{\partial \boldsymbol{\nu}} = -\frac{1}{3\lambda + 2\mu}\mathbb{C}\boldsymbol{n} \quad \text{on } \partial\Omega, \tag{3.139}$$

with the decay conditions at infinity

$$|\boldsymbol{\theta}^{qr}| = O(|\boldsymbol{x}|^{-1}), \qquad |\nabla\boldsymbol{\theta}^{qr}| = O(|\boldsymbol{x}|^{-2}), \qquad \text{as } |\boldsymbol{x}| \to \infty, \tag{3.140}$$

where the condition $\partial\boldsymbol{\theta}^{qr}/\partial\boldsymbol{\nu}$ has to be read as

$$\left(\frac{\partial \boldsymbol{\theta}^{qr}}{\partial \boldsymbol{\nu}}\right)_i = -\frac{1}{3\lambda + 2\mu}\mathbb{C}_{ijqr} n_j.$$

Well-posedness of this boundary value problem can be proved by integral equations or in weighted Sobolev spaces as did in the previous chapter.

With all these ingredients, we can state and prove the result related to the asymptotic expansion at first order of (3.137).

Theorem 3.4.1 (Asymptotic Expansion) *There exist $\varepsilon_0 > 0$ and a uniform constant $C > 0$ such that for all $\varepsilon \in (0, \varepsilon_0)$ the following expansion holds*

$$u_\varepsilon^k(\boldsymbol{y}) = \varepsilon^3|\Omega| p \widehat{\nabla}_z \boldsymbol{N}^{(k)}(z, \boldsymbol{y}) : \mathbb{M}\mathbf{I} + O(\varepsilon^4), \qquad \forall \boldsymbol{y} \in \mathbb{R}^2 \tag{3.141}$$

for $k = 1, 2, 3$, where $O(\varepsilon^4)$ denotes a quantity bounded by $C\varepsilon^4$, and $\mathbb{M}$ is the fourth-order moment elastic tensor defined by

$$\mathbb{M} := \mathbb{I} + \frac{1}{|\Omega|}\int_{\partial\Omega} \mathbb{C}(\boldsymbol{\theta}^{qr}(\boldsymbol{\zeta}) \otimes \boldsymbol{n}(\boldsymbol{\zeta}))\, d\sigma(\boldsymbol{\zeta}), \tag{3.142}$$

where $\boldsymbol{\theta}^{qr}$, for $q, r = 1, 2, 3$, are solutions of (3.139) *and* (3.140).

For the proof of this theorem, we first need to find the asymptotic behavior of the integral operators in (3.138).

Lemma 3.4.2 *For $\boldsymbol{x} = z + \varepsilon\boldsymbol{\zeta}$, with $\boldsymbol{\zeta} \in \partial\Omega$, the integral equation* (3.138) *is such that*

$$\left(\tfrac{1}{2}\mathbf{I} + \mathbf{K} + \varepsilon^2\boldsymbol{\Lambda}_{\Omega,\varepsilon}\right) \boldsymbol{f}^\sharp(\boldsymbol{\zeta}) = \varepsilon p \mathbf{S}^\Gamma(\boldsymbol{n})(\boldsymbol{\zeta}) + O(\varepsilon^2), \tag{3.143}$$

where

$$\boldsymbol{\Lambda}_{\Omega,\varepsilon} \boldsymbol{f}^\sharp(\boldsymbol{\eta}) := \int_{\partial\Omega} \frac{\partial \mathbf{R}}{\partial \boldsymbol{\nu}(\boldsymbol{\eta})}(z + \varepsilon\boldsymbol{\eta}, z + \varepsilon\boldsymbol{\zeta}) \boldsymbol{f}^\sharp(\boldsymbol{\eta})\, d\sigma(\boldsymbol{\eta})$$

is uniformly bounded in ε. Moreover, when ε is sufficiently small, we have

$$\boldsymbol{f}^\sharp(\boldsymbol{\zeta}) = \varepsilon p \left(\tfrac{1}{2}\mathbf{I} + \mathbf{K}\right)^{-1} \mathbf{S}^\Gamma(\boldsymbol{n})(\boldsymbol{\zeta}) + O(\varepsilon^2), \qquad \boldsymbol{\zeta} \in \partial\Omega. \tag{3.144}$$

Proof For $z + \varepsilon\boldsymbol{\zeta}$, where $\boldsymbol{\zeta} \in \partial\Omega$, we yield

$$\begin{aligned}\mathbf{D}_\varepsilon^R \boldsymbol{f}(z + \varepsilon\boldsymbol{\zeta}) &= \int_{\partial C_\varepsilon} \frac{\partial \mathbf{R}}{\partial \boldsymbol{\nu}(\boldsymbol{t})}(\boldsymbol{t}, z + \varepsilon\boldsymbol{\zeta}) \boldsymbol{f}(\boldsymbol{t})\, d\sigma(\boldsymbol{t}) \\ &= \varepsilon^2 \int_{\partial\Omega} \frac{\partial \mathbf{R}}{\partial \boldsymbol{\nu}(\boldsymbol{\eta})}(z + \varepsilon\boldsymbol{\eta}, z + \varepsilon\boldsymbol{\zeta}) \boldsymbol{f}^\sharp(\boldsymbol{\eta})\, d\sigma(\boldsymbol{\eta}).\end{aligned}$$

Since $\mathbf{R}$ is the regular part of the Neumann function (see Theorem 3.1.7), i.e. the kernel $\partial\mathbf{R}/\partial\boldsymbol{\nu}(\boldsymbol{\eta})$ is continuous, we get

$$\mathbf{D}_\varepsilon^R = \varepsilon^2 \boldsymbol{\Lambda}_{\Omega,\varepsilon} \tag{3.145}$$

where

$$\|\boldsymbol{\Lambda}_{\Omega,\varepsilon}\| \le C, \tag{3.146}$$

with C is uniform with respect to ε.

The integral

$$\mathbf{K}_\varepsilon \boldsymbol{f}(z + \varepsilon\boldsymbol{\zeta}) = \text{p.v.} \int_{\partial C_\varepsilon} \frac{\partial \boldsymbol{\Gamma}}{\partial \boldsymbol{\nu}(\boldsymbol{t})}(\boldsymbol{t} - z - \varepsilon\boldsymbol{\zeta}) \boldsymbol{f}(\boldsymbol{t})\, d\sigma(\boldsymbol{t})$$

can be handled by utilizing the explicit expression of the conormal derivative of the fundamental solution of the Lamé operator given in (3.37). In fact, making the substitution $\boldsymbol{t} = z+\varepsilon\boldsymbol{\eta}$ and recalling that (3.37) is a homogeneous function of degree -2, we get

$$\begin{aligned}
&\left(\frac{\partial \mathbf{\Gamma}^{(k)}}{\partial \boldsymbol{\nu}}\right)_h (\varepsilon(\boldsymbol{\eta}-\boldsymbol{\zeta}))\\
&\quad = -\frac{1}{4\pi\varepsilon^2}\left\{\left[\frac{1-2\nu}{2(1-\nu)}\delta_{hk} + \frac{3}{2(1-\nu)}\frac{\eta_k-\zeta_k}{|\boldsymbol{\eta}-\boldsymbol{\zeta}|}\frac{\eta_h-\zeta_h}{|\boldsymbol{\eta}-\boldsymbol{\zeta}|}\right]\frac{\partial}{\partial \boldsymbol{n}(\boldsymbol{\eta})}\frac{1}{|\boldsymbol{\eta}-\boldsymbol{\zeta}|}\right.\\
&\qquad\left. + \frac{1-2\nu}{2(1-\nu)}n_h(\boldsymbol{\eta})\frac{\eta_k-\zeta_k}{|\boldsymbol{\eta}-\boldsymbol{\zeta}|^3} - \frac{1-2\nu}{2(1-\nu)}n_k(\boldsymbol{\eta})\frac{\eta_h-\zeta_h}{|\boldsymbol{\eta}-\boldsymbol{\zeta}|^3}\right\}\\
&\quad = \frac{1}{\varepsilon^2}\left(\frac{\partial \mathbf{\Gamma}^{(k)}}{\partial \boldsymbol{\nu}}\right)_h (\boldsymbol{\eta}-\boldsymbol{\zeta}),
\end{aligned}$$

for $h, k = 1, 2, 3$. Therefore, it is straightforward to see that

$$\mathbf{K}_\varepsilon \boldsymbol{f}(z+\varepsilon\boldsymbol{\zeta}) = \text{p.v.}\int_{\partial\Omega} \frac{\partial \mathbf{\Gamma}}{\partial \boldsymbol{\nu}(\boldsymbol{\eta})}(\boldsymbol{\eta}-\boldsymbol{\zeta})\boldsymbol{f}^\sharp(\boldsymbol{\eta})\,d\sigma(\boldsymbol{\eta}) = \mathbf{K}\boldsymbol{f}^\sharp(\boldsymbol{\zeta}). \tag{3.147}$$

Again, by choosing $\boldsymbol{t} = z+\varepsilon\boldsymbol{\eta}$, with $\boldsymbol{\eta} \in \partial\Omega$, we find

$$\mathbf{S}_\varepsilon^\Gamma(\boldsymbol{n})(z+\varepsilon\boldsymbol{\zeta}) = \varepsilon^2\int_{\partial\Omega} \mathbf{\Gamma}(\varepsilon(\boldsymbol{\eta}-\boldsymbol{\zeta}))\boldsymbol{n}(\boldsymbol{\eta})\,d\sigma(\boldsymbol{\eta}) = \varepsilon\mathbf{S}^\Gamma(\boldsymbol{n})(\boldsymbol{\zeta}), \tag{3.148}$$

where the last equality follows from the fact that the fundamental solution is homogeneous of degree -1. Similarly

$$\mathbf{S}_\varepsilon^R(\boldsymbol{n})(z+\varepsilon\boldsymbol{\zeta}) = \varepsilon^2\int_{\partial\Omega} \mathbf{R}(z+\varepsilon\boldsymbol{\eta}, z+\varepsilon\boldsymbol{\zeta})\boldsymbol{n}(\boldsymbol{\eta})\,d\sigma(\boldsymbol{\eta}),$$

and since $\mathbf{R}$ is regular, see Theorem 3.1.7, it follows that

$$\mathbf{S}_\varepsilon^R(\boldsymbol{n})(z+\varepsilon\boldsymbol{\zeta}) = O(\varepsilon^2). \tag{3.149}$$

The first part of the statement of the theorem, i.e. Eq. (3.143), follows putting together all the previous result, that is (3.145) and (3.147)–(3.149).

The second part of the statement of the theorem follows by noticing that from (3.143)

$$\left(\tfrac{1}{2}\mathbf{I}+\mathbf{K}\right)\left(\mathbf{I}+\varepsilon^2\left(\tfrac{1}{2}\mathbf{I}+\mathbf{K}\right)^{-1}\mathbf{\Lambda}_{\varepsilon,\Omega}\right)\boldsymbol{f}^\sharp = \varepsilon p\mathbf{S}^\Gamma(\boldsymbol{n}) + O(\varepsilon^2), \qquad \text{on } \partial\Omega.$$

Therefore, from Lemma 3.2.6 and since the operator $\mathbf{\Lambda}_{\varepsilon,\Omega}$ is continuous, see (3.146), we yield

$$\left\| \left(\tfrac{1}{2}\mathbf{I}+\mathbf{K}\right)^{-1} \mathbf{\Lambda}_{\varepsilon,\Omega} \right\| \leq C,$$

where $C > 0$ is independent from ε. On the other hand, choosing $\varepsilon_0^2 = 1/C$, it follows that for all $\varepsilon \in (0, \varepsilon_0)$ the operator

$$\mathbf{I} + \varepsilon^2 \left(\tfrac{1}{2}\mathbf{I}+\mathbf{K}\right)^{-1} \mathbf{\Lambda}_{\varepsilon,\Omega}$$

is invertible. Therefore by utilizing the Neumann series, which we truncate at zero order, we find

$$\left(\mathbf{I} + \varepsilon^2 \left(\tfrac{1}{2}\mathbf{I}+\mathbf{K}\right)^{-1} \mathbf{\Lambda}_{\varepsilon,\Omega}\right)^{-1} = \mathbf{I} + O(\varepsilon^2).$$

Therefore

$$\boldsymbol{f}^{\sharp} = \varepsilon p \left(\tfrac{1}{2}\mathbf{I}+\mathbf{K}\right)^{-1} \mathbf{S}^{\Gamma}(\boldsymbol{n}) + O(\varepsilon^2), \qquad \text{on } \partial\Omega,$$

that is the second part of the assertion of the theorem. □

We can interpret the term $\left(\tfrac{1}{2}\mathbf{I}+\mathbf{K}\right)^{-1} \mathbf{S}^{\Gamma}(\boldsymbol{n})(\boldsymbol{\zeta})$ in (3.144) as the trace on the boundary of Ω of the solution of an exterior problem in the whole space. To this end, we define the function $\boldsymbol{w} : \partial\Omega \to \partial\Omega$ as

$$\boldsymbol{w}(\boldsymbol{\zeta}) := -\left(\tfrac{1}{2}\mathbf{I}+\mathbf{K}\right)^{-1} \mathbf{S}^{\Gamma}(\boldsymbol{n})(\boldsymbol{\zeta}), \qquad \boldsymbol{\zeta} \in \partial\Omega. \tag{3.150}$$

and we consider the problem

$$\operatorname{div}\left(\mathbb{C}\widehat{\nabla}\boldsymbol{v}\right) = 0 \qquad \text{in } \mathbb{R}^3 \setminus \Omega, \qquad \frac{\partial \boldsymbol{v}}{\partial \boldsymbol{\nu}} = -\boldsymbol{n} \qquad \text{on } \partial\Omega \tag{3.151}$$

with decay conditions at infinity

$$\boldsymbol{v} = O(|\boldsymbol{x}|^{-1}), \qquad |\nabla \boldsymbol{v}| = O(|\boldsymbol{x}|^{-2}) \qquad \text{as } |\boldsymbol{x}| \to +\infty. \tag{3.152}$$

As before, utilizing the weighted Sobolev spaces (without using the explicit decay at infinity (3.152)), introduced in Sect. 2.1.1, it is straightforward to prove the well-posedness of this boundary value problem on $\boldsymbol{v}$, see [30]. We skip this part and we prove the following result for $\boldsymbol{w}$.

Proposition 3.4.3 *The function* $\boldsymbol{w}$*, defined in* (3.150)*, is such that* $\boldsymbol{w} = \boldsymbol{v}\big|_{\boldsymbol{x}\in\partial\Omega}$ *where* $\boldsymbol{v}$ *is the solution to* (3.151) *and* (3.152).

Proof We follow a similar approach used for the proof of Theorem 3.2.1. To get an integral representation formula, we use the second Betti's formula, see (3.5), which is applied to the fundamental solution $\boldsymbol{\Gamma}$, see (3.6), and the function $\boldsymbol{v}$ into the domain $B_r(\boldsymbol{0}) \setminus (\Omega \cup B_\varepsilon(\boldsymbol{x}))$, where $\varepsilon > 0$ and $r > 0$ which is assumed to be sufficiently large such that to contain the cavity Ω. In this way, we find

$$\boldsymbol{v}(\boldsymbol{x}) = -\mathbf{S}^\Gamma \boldsymbol{n}(\boldsymbol{x}) - \mathbf{D}^\Gamma \boldsymbol{v}(\boldsymbol{x}), \qquad \boldsymbol{x} \in \mathbb{R}^3 \setminus \Omega$$

Therefore, from the single and double layer potential properties for the Lamé system, see (3.10), we find

$$\boldsymbol{v}(\boldsymbol{x}) = -\mathbf{S}^\Gamma \boldsymbol{n}(\boldsymbol{x}) - \left(-\tfrac{1}{2}\mathbf{I} + \mathbf{K}\right) \boldsymbol{v}(\boldsymbol{x}), \qquad \boldsymbol{x} \in \partial\Omega$$

hence

$$\boldsymbol{v}(\boldsymbol{x}) = -\left(\tfrac{1}{2}\mathbf{I} + \mathbf{K}\right)^{-1} \mathbf{S}^\Gamma(\boldsymbol{n})(\boldsymbol{x}), \qquad \boldsymbol{x} \in \partial\Omega$$

that is the assertion. □

Using the auxiliary functions $\boldsymbol{\theta}^{qr}$ defined in (3.139) and (3.140), we rewrite the function $\boldsymbol{v}$, and consequently its trace $\boldsymbol{w}$ on $\partial\Omega$, as

$$\boldsymbol{v} = \boldsymbol{\theta}^{qr}\delta_{qr},$$

where we use the convention to sum up the repeated indices. It is straightforward to see that the elastostatic equation and the boundary condition in (3.151) are satisfied.

Proof of Theorem 3.4.1 We find the asymptotic expansion of the two integrals $I_1^{(k)}, I_2^{(k)}$ defined in (3.137). Since $\boldsymbol{y} \in \mathbb{R}^2$ and $\boldsymbol{x} \in \partial C_\varepsilon = \boldsymbol{z} + \varepsilon\boldsymbol{\zeta}$, with $\boldsymbol{\zeta} \in \partial\Omega$, we consider the Taylor expansion of the Neumann function, i.e.

$$\boldsymbol{N}^{(k)}(\boldsymbol{z} + \varepsilon\boldsymbol{\zeta}, \boldsymbol{y}) = \boldsymbol{N}^{(k)}(\boldsymbol{z}, \boldsymbol{y}) + \varepsilon\nabla\boldsymbol{N}^{(k)}(\boldsymbol{z}, \boldsymbol{y})\boldsymbol{\zeta} + O(\varepsilon^2), \tag{3.153}$$

for $k = 1, 2, 3$. Utilizing this expansion and the change of variable $\boldsymbol{x} = \boldsymbol{z} + \varepsilon\boldsymbol{\zeta}$ in $I_1^{(k)}$, we yield

$$\begin{aligned} I_1^{(k)} &= \varepsilon^2 p \boldsymbol{N}^{(k)}(\boldsymbol{z}, \boldsymbol{y}) \cdot \int_{\partial\Omega} \boldsymbol{n}\, d\sigma(\boldsymbol{\zeta}) + \varepsilon^3 p \int_{\partial\Omega} \boldsymbol{n}(\boldsymbol{\zeta}) \cdot \nabla\boldsymbol{N}^{(k)}(\boldsymbol{z}, \boldsymbol{y})\boldsymbol{\zeta}\, d\sigma(\boldsymbol{\zeta}) + O(\varepsilon^4) \\ &:= p\left(\varepsilon^2 I_{11}^{(k)} + \varepsilon^3 I_{12}^{(k)}\right) + O(\varepsilon^4). \end{aligned}$$

Integral $I_{11}^{(k)}$ is null, in fact, as a consequence of the application of the divergence theorem

$$\int_{\partial\Omega} \boldsymbol{n}(\boldsymbol{\zeta})\, d\sigma(\boldsymbol{\zeta}) = \boldsymbol{0}.$$

We use the equality $\boldsymbol{n} \cdot \nabla \boldsymbol{N}^{(k)} \boldsymbol{\zeta} = \nabla \boldsymbol{N}^{(k)} : (\boldsymbol{n}(\boldsymbol{\zeta}) \otimes \boldsymbol{\zeta})$ in $I_{12}^{(k)}$ which gives

$$I_1^{(k)} = \varepsilon^3 p \nabla \boldsymbol{N}^{(k)}(z, \boldsymbol{y}) : \int_{\partial\Omega} \big(\boldsymbol{n}(\boldsymbol{\zeta}) \otimes \boldsymbol{\zeta}\big)\, d\sigma(\boldsymbol{\zeta}) + O(\varepsilon^4), \qquad k = 1, 2, 3. \tag{3.154}$$

In $I_2^{(k)}$ we apply the result in Lemma 3.4.2 and we use the Taylor expansion of the conormal derivative of $\mathbf{N}^{(k)}(\boldsymbol{x}, \boldsymbol{y})$, for $k = 1, 2, 3$ truncated at order zero, that is for $\boldsymbol{x} = z + \varepsilon\boldsymbol{\zeta}$, when $\boldsymbol{\zeta} \in \partial\Omega$ and $\boldsymbol{y} \in \mathbb{R}^2$,

$$\frac{\partial \boldsymbol{N}^{(k)}}{\partial \boldsymbol{\nu}(\boldsymbol{x})}(\boldsymbol{x}, \boldsymbol{y}) = \frac{\partial \boldsymbol{N}^{(k)}}{\partial \boldsymbol{\nu}(\boldsymbol{\zeta})}(z, \boldsymbol{y}) + O(\varepsilon), \qquad k = 1, 2, 3.$$

Therefore

$$\begin{aligned} I_2^{(k)} &= -\varepsilon^2 \int_{\partial\Omega} \frac{\partial \boldsymbol{N}^{(k)}}{\partial \boldsymbol{\nu}(\boldsymbol{x})}(z + \boldsymbol{\varepsilon\zeta}, \boldsymbol{y}) \cdot \boldsymbol{f}^{\sharp}(\boldsymbol{\zeta})\, d\sigma(\boldsymbol{\zeta}) \\ &= \varepsilon^3 p \int_{\partial\Omega} \frac{\partial \boldsymbol{N}^{(k)}}{\partial \boldsymbol{\nu}(\boldsymbol{\zeta})}(z, \boldsymbol{y}) \cdot \boldsymbol{w}(\boldsymbol{\zeta})\, d\sigma(\boldsymbol{\zeta}) + O(\varepsilon^4), \end{aligned}$$

for any k, where $\boldsymbol{w}$ is defined in (3.150). Since $\partial \boldsymbol{N}^{(k)}/\partial \boldsymbol{\nu}(\boldsymbol{\zeta}) = \mathbb{C}\widehat{\nabla} \boldsymbol{N}^{(k)} \boldsymbol{n}(\boldsymbol{\zeta})$, we have

$$\mathbb{C}\widehat{\nabla} \boldsymbol{N}^{(k)} \boldsymbol{n}(\boldsymbol{\zeta}) \cdot \boldsymbol{w}(\boldsymbol{\zeta}) = \mathbb{C}\widehat{\nabla} \boldsymbol{N}^{(k)} : (\boldsymbol{w}(\boldsymbol{\zeta}) \otimes \boldsymbol{n}(\boldsymbol{\zeta})).$$

Therefore

$$I_2^{(k)}(\boldsymbol{y}) = \varepsilon^3\, p\, \mathbb{C}\widehat{\nabla} \boldsymbol{N}^{(k)}(z, \boldsymbol{y}) : \int_{\partial\Omega} (\boldsymbol{w}(\boldsymbol{\zeta}) \otimes \boldsymbol{n}(\boldsymbol{\zeta}))\, d\sigma(\boldsymbol{\zeta}) + O(\varepsilon^4). \tag{3.155}$$

Collecting the result in (3.154) and (3.155), Eq. (3.137) gives

$$\begin{aligned} u_\varepsilon^k(\boldsymbol{y}) &= I_1^{(k)}(\boldsymbol{y}) + I_2^{(k)}(\boldsymbol{y}) \\ &= \varepsilon^3 p \Big[\nabla \boldsymbol{N}^{(k)}(z, \boldsymbol{y}) : \int_{\partial\Omega} (\boldsymbol{n} \otimes \boldsymbol{\zeta})\, d\sigma + \mathbb{C}\widehat{\nabla} \boldsymbol{N}^{(k)}(z, \boldsymbol{y}) : \int_{\partial\Omega} (\boldsymbol{w} \otimes \boldsymbol{n})\, d\sigma \Big] + O(\varepsilon^4). \end{aligned}$$

Now, handling this expression, we highlight the moment elastic tensor. We have

$$\int_{\partial\Omega} (\boldsymbol{n}(\boldsymbol{\zeta}) \otimes \boldsymbol{\zeta})\, d\sigma(\boldsymbol{\zeta}) = |\Omega|\mathbf{I}, \tag{3.156}$$

indeed, for any $i, j = 1, 2, 3$, it follows

$$\begin{aligned}\int_{\partial\Omega} \zeta_i n_j \, d\sigma(\boldsymbol{\zeta}) &= \int_{\partial\Omega} \boldsymbol{n} \cdot \zeta_i \boldsymbol{e}_j \, d\sigma(\boldsymbol{\zeta}) \\ &= \int_{\Omega} \operatorname{div}\left(\zeta_i \boldsymbol{e}_j\right) d\boldsymbol{\zeta} = \int_{\Omega} \boldsymbol{e}_j \cdot \boldsymbol{e}_i \, d\boldsymbol{\zeta} = |\Omega|\delta_{ij},\end{aligned}$$

where $\boldsymbol{e}_j$ is the j-th unit vector of $\mathbb{R}^3$. Hence, by (3.156) and taking the symmetric part of $\nabla \boldsymbol{N}^{(k)}$, for any k, we find

$$u_\varepsilon^k = \varepsilon^3 p\left[\widehat{\nabla}\boldsymbol{N}^{(k)} : \mathbf{I}|\Omega| + \mathbb{C}\widehat{\nabla}\boldsymbol{N}^{(k)} : \int_{\partial\Omega} \boldsymbol{w} \otimes \boldsymbol{n}\, d\sigma(\boldsymbol{\zeta})\right] + O(\varepsilon^4).$$

From the symmetries of elasticity tensor $\mathbb{C}$, we have

$$u_\varepsilon^k = \varepsilon^3 |\Omega| p \widehat{\nabla}\boldsymbol{N}^{(k)} : \left[\mathbf{I} + \frac{1}{|\Omega|}\int_{\partial\Omega} \mathbb{C}(\boldsymbol{w} \otimes \boldsymbol{n})\, d\sigma(\boldsymbol{\zeta})\right] + O(\varepsilon^4),$$

for $k = 1, 2, 3$. The assertion of the theorem follows by noticing that $\mathbf{I} = \mathbb{I}$ and $\boldsymbol{w} = \boldsymbol{\theta}^{qr}\delta_{qr}$. □

3.4.1 *Properties of the Moment Elastic Tensor*

For a complete analysis, in this section we find all the symmetries satisfied by the fourth-order tensor $\mathbb{M}$ and we study its positivity. We apply similar ideas contained in [106].

For this purpose, we use the equations (3.139) and in particular their weak formulation in $H_w^1(\mathbb{R}^3 \setminus \Omega)$. Let us assume, for the moment, $\boldsymbol{\theta}^{kh}$ regular. For all $\boldsymbol{\varphi} \in C_0^\infty(\mathbb{R}^3 \setminus \Omega)$, we find

$$\begin{aligned}0 &= \int_{\mathbb{R}^3\setminus\Omega} \operatorname{div}(\mathbb{C}\widehat{\nabla}\boldsymbol{\theta}^{kh}) \cdot \boldsymbol{\varphi}\, d\boldsymbol{x} \\ &= -\int_{\partial\Omega} (\mathbb{C}\widehat{\nabla}\boldsymbol{\theta}^{kh}\boldsymbol{n}) \cdot \boldsymbol{\varphi}\, d\sigma(\boldsymbol{x}) - \int_{\mathbb{R}^3\setminus\Omega} \mathbb{C}\widehat{\nabla}\boldsymbol{\theta}^{kh} : \widehat{\nabla}\boldsymbol{\varphi}\, d\boldsymbol{x},\end{aligned}$$

hence

$$\int_{\mathbb{R}^3\setminus\Omega} \mathbb{C}\widehat{\nabla}\boldsymbol{\theta}^{kh} : \widehat{\nabla}\boldsymbol{\varphi}\, d\boldsymbol{x} = -\int_{\partial\Omega} (\mathbb{C}\widehat{\nabla}\boldsymbol{\theta}^{kh}\boldsymbol{n}) \cdot \boldsymbol{\varphi}\, d\sigma(\boldsymbol{x}).$$

Now, utilizing the boundary conditions in (3.139), we get

$$\int_{\mathbb{R}^3\setminus\Omega} \mathbb{C}\widehat{\nabla}\boldsymbol{\theta}^{kh} : \widehat{\nabla}\boldsymbol{\varphi}\, d\boldsymbol{x} = \frac{1}{3\lambda+2\mu}\int_{\partial\Omega} \mathbb{C}(\boldsymbol{n}\otimes\boldsymbol{\varphi})\, d\sigma(\boldsymbol{x}).$$

For the density result of $C_0^\infty(\mathbb{R}^3 \setminus \Omega)$ in $H^1_w(\mathbb{R}^3 \setminus \Omega)$, the previous equation is also true in $H^1_w(\mathbb{R}^3 \setminus \Omega)$, see [30]. Choosing $\boldsymbol{\varphi} = \boldsymbol{\theta}^{rs}$, with $r, s = 1, 2, 3$, we have

$$\int_{\mathbb{R}^3\setminus\Omega} \mathbb{C}\widehat{\nabla}\boldsymbol{\theta}^{kh} : \widehat{\nabla}\boldsymbol{\theta}^{rs}\, d\boldsymbol{x} = \frac{1}{3\lambda+2\mu}\int_{\partial\Omega} \mathbb{C}(\boldsymbol{n}\otimes\boldsymbol{\theta}^{rs})\, d\sigma(\boldsymbol{x}) \tag{3.157}$$

or in components, summing up the repeated indices,

$$\int_{\mathbb{R}^3\setminus\Omega} \mathbb{C}_{ijlm}(\widehat{\nabla}\boldsymbol{\theta}^{kh})_{lm}(\widehat{\nabla}\boldsymbol{\theta}^{rs})_{ij}\, d\boldsymbol{x} = \frac{1}{3\lambda+2\mu}\int_{\partial\Omega} \mathbb{C}_{ijkh}\, n_i\theta^{rs}_j\, d\sigma(\boldsymbol{x}).$$

Positivity

Now, we prove the positivity of the tensor $\mathbb{M}$, i.e. $\mathbb{M}\mathbf{A} : \mathbf{A} > 0$, for all $\mathbf{A} \in \mathbb{R}^{3\times 3}$. By the definition (3.142) of $\mathbb{M}$ and applying (3.157) we have

$$\begin{aligned}\mathbb{M}_{khrs}A_{kh}A_{rs} &= |\mathbf{A}|^2 + (3\lambda+2\mu)\int_{\mathbb{R}^3\setminus\Omega} \mathbb{C}\widehat{\nabla}\boldsymbol{\theta}^{kh} : \widehat{\nabla}\boldsymbol{\theta}^{rs}A_{kh}A_{rs}\, d\boldsymbol{x} \\ &= |\mathbf{A}|^2 + (3\lambda+2\mu)\int_{\mathbb{R}^3\setminus\Omega} \mathbb{C}\widehat{\nabla}(\boldsymbol{\theta}^{kh}A_{kh}) : (\widehat{\nabla}\boldsymbol{\theta}^{rs}A_{rs})\, d\boldsymbol{x} > 0\end{aligned}$$

since $\mathbb{C}$ is positive definite.

Symmetries

First, we notice that from $\boldsymbol{w} = \boldsymbol{\theta}^{kh}\delta_{kh}$ we have $\boldsymbol{w} = \boldsymbol{\theta}^{hk}\delta_{hk}$, hence $\boldsymbol{\theta}^{hk}$ satisfy the same problem (3.139) and (3.140). Again, by the definition (3.142), the weak formulation (3.157) and the symmetries of the elastic tensor $\mathbb{C}$, it is straightforward to obtain the following symmetries for the moment elastic tensor

$$\mathbb{M}_{khrs} = \mathbb{M}_{hkrs} = \mathbb{M}_{khsr} = \mathbb{M}_{rskh},$$

where $k, h, r, s = 1, 2, 3$.

3.4.2 *The Mogi Model*

In this section we specialized the explicit asymptotic expansion (3.141) to the case of spherical cavities which, from an application point of view, corresponds to the Mogi model, see Chap. 1 for a more in-depth discussion. We first recall the asymptotic expansion (3.141), i.e.,

$$u_\varepsilon^k(\boldsymbol{y}) = \varepsilon^3 |\Omega| p \widehat{\nabla}_z \boldsymbol{N}^{(k)}(z, \boldsymbol{y}) : \mathbb{M}\mathbf{I} + O(\varepsilon^4), \qquad k = 1, 2, 3, \tag{3.158}$$

where $\mathbb{M}$ is the tensor given in (3.142). We first note that

$$\mathbb{M}\mathbf{I} = \left[\mathbb{I} + \frac{1}{|\Omega|} \int_{\partial\Omega} \mathbb{C}(\boldsymbol{\theta}^{qr} \otimes \boldsymbol{n})\, d\sigma(\boldsymbol{\zeta}) \right] \mathbf{I} = \mathbf{I} + \frac{1}{|\Omega|} \int_{\partial\Omega} \mathbb{C}(\boldsymbol{w} \otimes \boldsymbol{n})\, d\sigma(\boldsymbol{\zeta}), \tag{3.159}$$

where, in the last equality, we use the connection between the functions $\boldsymbol{w}$ and $\boldsymbol{\theta}^{qr}$ on $\partial\Omega$ that is $\boldsymbol{w} = \boldsymbol{\theta}^{qr}\delta_{qr}$, $q, r = 1, 2, 3$. The first step is to find the explicit expression of $\boldsymbol{w}$, trace on $\partial\Omega$ of the solution $\boldsymbol{v}$ to (3.151), when the cavity is the unit sphere. Then we calculate the gradient of the Neumann function $\mathbf{N}$.

We recall the problem for $\boldsymbol{v}$ in the case where the cavity is the unit sphere, i.e.,

$$\operatorname{div}(\mathbb{C}\widehat{\nabla}\boldsymbol{v}) = 0 \quad \text{in } \mathbb{R}^3 \setminus B_1(\mathbf{0}), \qquad \frac{\partial \boldsymbol{v}}{\partial \nu} = -\boldsymbol{n} \quad \text{on } \partial B_1(\mathbf{0}),$$

where $B_1(\mathbf{0}) = \{\boldsymbol{x} \in \mathbb{R}^3 : |\boldsymbol{x}| \le 1\}$ with decay at infinity

$$\boldsymbol{v} = O(|\boldsymbol{x}|^{-1}), \qquad |\nabla \boldsymbol{v}| = O(|\boldsymbol{x}|^{-2}) \qquad \text{as } |\boldsymbol{x}| \to +\infty.$$

We look for a solution with the form

$$\boldsymbol{v}(\boldsymbol{x}) = \phi(r)\,\boldsymbol{x} \qquad \text{with } r := |\boldsymbol{x}|,$$

so that

$$\Delta v_i = \left\{ \phi'' + \frac{4\phi'}{r} \right\} x_i, \qquad \operatorname{div} \boldsymbol{v} = r\phi' + 3\phi, \qquad \nabla \operatorname{div} \boldsymbol{v} = \left\{ \phi'' + \frac{4\phi'}{r} \right\} \boldsymbol{x}.$$

Since $\boldsymbol{n} = \boldsymbol{x}$ on ∂B, by direct calculations, we get

$$\operatorname{div}(\mathbb{C}\widehat{\nabla}\boldsymbol{v}) = (\lambda + 2\mu) \left(\phi'' + \frac{4\phi'}{r} \right) \boldsymbol{x},$$

$$\frac{\partial \boldsymbol{v}}{\partial \nu} = \big\{ (\lambda + 2\mu) r\phi' + (3\lambda + 2\mu)\phi \big\} \boldsymbol{x}$$

Thefore, we need to find a function $\phi : [1, +\infty) \to \mathbb{R}$ such that

$$\phi'' + \frac{4\phi'}{r} = 0, \quad (\lambda + 2\mu)r\phi' + (3\lambda + 2\mu)\phi\Big|_{r=1} = -1, \quad \phi\Big|_{r=+\infty} = 0.$$

By means of the decay conditions at infinity, we find that $B = 0$ and $A = 1/4\mu$. Therefore, the solution is $\mathbf{v}(\boldsymbol{x}) = \boldsymbol{x}/4\mu|\boldsymbol{x}|^3$, which implies that

$$\boldsymbol{w}(\boldsymbol{x}) := \boldsymbol{v}(\boldsymbol{x})\Big|_{|\boldsymbol{x}|=1} = \frac{\boldsymbol{x}}{4\mu}.$$

Substituting this expression of $\boldsymbol{w}$ in (3.159) (where now $\Omega = B_1(\mathbf{0})$), we get

$$\begin{aligned}\mathbb{MI} &= \mathbf{I} + \frac{1}{|B_1(\mathbf{0})|} \int\limits_{\partial B_1(\mathbf{0})} \mathbb{C}(\boldsymbol{w}(\boldsymbol{\zeta}) \otimes \boldsymbol{n}(\boldsymbol{\zeta}))\, d\sigma(\zeta) \\ &= \mathbf{I} + \frac{3}{16\pi\mu} \int\limits_{\partial B_1(\mathbf{0})} \frac{\mathbb{C}(\boldsymbol{\zeta} \otimes \boldsymbol{\zeta})}{|\boldsymbol{\zeta}|^3}\, d\sigma(\boldsymbol{\zeta}).\end{aligned}$$

By means of the spherical coordinates and using the orthogonality conditions for the circular functions, it holds

$$\int\limits_{\partial B_1(\mathbf{0})} \frac{\boldsymbol{\zeta} \otimes \boldsymbol{\zeta}}{|\boldsymbol{\zeta}|^3}\, d\sigma(\boldsymbol{\zeta}) = \frac{4\pi}{3}\mathbf{I},$$

hence the second-order tensor $\mathbb{MI}$ is given by

$$\mathbb{MI} = \frac{3(\lambda + 2\mu)}{4\mu}\mathbf{I}.$$

Inserting this result in (3.158), we find

$$u_\varepsilon^k(\boldsymbol{y}) = \frac{\pi(\lambda + 2\mu)}{\mu}\varepsilon^3 p\, \mathrm{Tr}(\widehat{\nabla}_z \boldsymbol{N}^{(k)}(z, \boldsymbol{y})) + O(\varepsilon^4), \qquad k = 1, 2, 3. \tag{3.160}$$

We have now to evaluate the gradient of the Neumann function $\mathbf{N}$ at $y_3 = 0$, see Theorem 3.1.7 for the explicit components of the matrix $\mathbf{N}$.

For $y_3 = 0$ we have that $\mathbf{N} = \mathbf{N}(z, \boldsymbol{y})$ is equal to

$$\begin{aligned}\kappa_\mu^{-1} N_{\alpha\alpha} &= -f - (z_\alpha - y_\alpha)^2 f^3 - (1-2\nu)g + (1-2\nu)(z_\alpha - y_\alpha)^2 f g^2 \\ \kappa_\mu^{-1} N_{\beta\alpha} &= (z_\alpha - y_\alpha)(z_\beta - y_\beta)\{-f^3 + (1-2\nu) f g\} \\ \kappa_\mu^{-1} N_{3\alpha} &= (z_\alpha - y_\alpha)\{-z_3 f^3 + (1-2\nu) f g\} \\ \kappa_\mu^{-1} N_{\alpha 3} &= (z_\alpha - y_\alpha)\Big\{-z_3 f^3 - (1-2\nu) f g\Big\} \\ \kappa_\mu^{-1} N_{33} &= -2(1-\nu) f - z_3^2 f^3,\end{aligned}$$

where $\alpha, \beta = 1, 2$ and $\kappa_\mu = 1/(4\pi\mu)$, with

$$f = 1/|z - \mathbf{y}|, \qquad \text{and} \qquad g = 1/(|z - \mathbf{y}| - z_3).$$

Let $\rho^2 := (z_1 - y_1)^2 + (z_2 - y_2)^2$. Using the identities

$$\rho^2 f^2 = 1 - z_3^2 f^2, \qquad (1 - z_3 f)g = f$$

and the differentiation formulas

$$\begin{aligned}
\partial_{z_\alpha} f &= -(z_\alpha - y_\alpha) f^3, & \partial_{z_3} f &= -z_3 f^3\\
\partial_{z_\alpha} g &= -(z_\alpha - y_\alpha) fg, & \partial_{z_3} g &= fg,\\
\partial_{z_\alpha}(fg) &= -(z_\alpha - y_\alpha)(f + g) f^2 g, & \partial_{z_3}(fg) &= f^3,
\end{aligned}$$

we derive the following formulas for some of the derivatives of $\kappa_\mu^{-1} N_{ij}$

$$\begin{aligned}
\kappa_\mu^{-1}\partial_{z_\alpha} N_{\alpha\alpha} &= (z_\alpha - y_\alpha)\big\{ - f^3 + 3(z_\alpha - y_\alpha)^2 f^5\\
&\quad + (1 - 2\nu)\big[3f - (z_\alpha - y_\alpha)^2 f^2 (f + 2g)\big] g^2\big\}\\
\kappa_\mu^{-1}\partial_{z_\beta} N_{\beta\alpha} &= (z_\alpha - y_\alpha)\big\{ - f^3 + 3(z_\beta - y_\beta)^2 f^5\\
&\quad + (1 - 2\nu)\big[f - (z_\beta - y_\beta)^2 f^2 (f + 2g)\big] g^2\big\}\\
\kappa_\mu^{-1}\partial_{z_3} N_{3\alpha} &= (z_\alpha - y_\alpha)\big\{ - 2\nu f^3 + 3z_3^2 f^5\big\}\\
\kappa_\mu^{-1}\partial_{z_\alpha} N_{\alpha 3} &= -z_3 f^3 + 3(z_\alpha - y_\alpha)^2 z_3 f^5\\
&\quad + (1 - 2\nu)\big[- 1 + (z_\alpha - y_\alpha)^2 (f + g) f\big] fg\\
\kappa_\mu^{-1}\partial_{z_3} N_{33} &= -2\nu z_3 f^3 + 3 z_3^3 f^5.
\end{aligned}$$

Therefore, we yield

$$\begin{aligned}
\mathrm{Tr}\big(\hat{\nabla} \boldsymbol{N}^{(\alpha)}\big) &= 2\kappa_\mu (1 - 2\nu)(z_\alpha - y_\alpha) f^3, \qquad \text{for } \alpha = 1, 2\\
\mathrm{Tr}\big(\hat{\nabla} \boldsymbol{N}^{(3)}\big) &= 2\kappa_\mu (1 - 2\nu) z_3 f^3.
\end{aligned} \tag{3.161}$$

Combining (3.160), (3.161) and using the explicit expression for f, we find

$$\begin{aligned}
u_\varepsilon^\alpha(\mathbf{y}) &= \frac{1 - \nu}{\mu} \frac{\varepsilon^3 p (z_\alpha - y_\alpha)}{|z - \mathbf{y}|^3} + O(\varepsilon^4), \qquad \text{for } \alpha = 1, 2\\
u_\varepsilon^3(\mathbf{y}) &= \frac{1 - \nu}{\mu} \frac{\varepsilon^3 p\, z_3}{|z - \mathbf{y}|^3} + O(\varepsilon^4),
\end{aligned}$$

that are the components given in (1.1).

We highlight that, in general, for other shapes of the cavity Ω, the trace on $\partial\Omega$ of the auxiliary functions $\boldsymbol{\theta}^{qr}$, with $q, r = 1, 2, 3$, can be numerically approximated (if it cannot be calculated explicitly) and, thus, the first term in the asymptotic expansion (3.141) can be considered as known in practical cases.

Bibliography

1. Abramovich, Y.A., Aliprantis, C.D.: An Invitation to Operator Theory. Graduate Studies in Mathematics, vol. 50. American Mathematical Society, Providence (2002)
2. Agmon, S.: Unicité et convexité dans les problèmes differéntiels. Sém. de Mathématiques Sup., Univ. de Montréal (1965)
3. Alessandrini, G.: Examples of instability in inverse boundary-value problems. Inverse Prob. **13**, 887–897 (1997)
4. Alessandrini, G., Morassi, A.: Strong unique continuation for Lamé system of elasticity. Commun. Partial Differ. Equ. **26**, 1787–1810 (2001)
5. Alessandrini, G., Rondi, L.: Optimal stability for the inverse problem of multiple cavities. J. Differ. Equ. **176**, 356–386 (2001)
6. Alessandrini, G., Sincich, E.: Cracks with impedance; stable determination from boundary data. Indiana Univ. Math. J. **62**, 947–989 (2013)
7. Alessandrini, G., Beretta, E., Rosset, E., Vessella, S.: Optimal stability for inverse elliptic boundary value problems with unknown boundaries. Ann. Scuola Norm. Sup. Pisa Cl. Sci. **29**, 755–806 (2000)
8. Alessandrini, G., Rosset, E., Seo, J.K.: Optimal size estimates for the inverse conductivity problem with one measurement. Proc. Am. Math. Soc. **128**, 53–64 (2000)
9. Alessandrini, G., Morassi, A., Rosset, E.: Detecting cavities by electrostatic boundary measurements. Inverse Prob. **18**, 1333–1353 (2002)
10. Alessandrini, G., Morassi, A., Rosset, E.: Detecting an inclusion in an elastic body by boundary measurements. SIAM J. Math. Anal. **33**, 1247–1268 (2002)
11. Alessandrini, G., Morassi, A., Rosset, E.: Size estimates. In: Alessandrini, G., Uhlmann, G. (eds.) Inverse Problems: Theory and Applications. Contemporary Mathematics, vol. 333, pp. 1–33. American Mathematical Society, Providence (2003)
12. Alessandrini, G., Morassi, A., Rosset, E.: The linear constraints in Poincaré and Korn type inequalities. Forum Math **20**, 557–569 (2008)
13. Alessandrini, G., Morassi, A., Rosset, E., Vessella, S.: On doubling inequalities for elliptic systems. J. Math. Anal. Appl. **357**, 349–355 (2009)
14. Alessandrini, G., Rondi, L., Rosset, E., Vessella, S.: The stability for the Cauchy problem for elliptic equations. Inverse Prob. **25**, 123004 (47 pp.) (2009)
15. Alessandrini, G., Di Cristo, M., Morassi, A., Rosset, E.: Stable determination of an inclusion in an elastic body by boundary measurements. SIAM J. Math. Anal. **46**, 2692–2729 (2014)
16. Alessandrini, G., Morassi, A., Rosset, E., Vessella, S.: Global stability for an inverse problem in soil-structure interaction. Proc. R. Soc. A **471**, 20150117 (12 pp.) (2015)

A. Aspri, *An Elastic Model for Volcanology*, Lecture Notes in Geosystems Mathematics and Computing, https://doi.org/10.1007/978-3-030-31475-0

17. Alves, C., Ammari, H.: Boundary integral formulae for the reconstruction of imperfections of small diameter in an elastic medium. SIAM J. Appl. Math. **62**, 94–106 (2001)
18. Ammari, H.: An Introduction to Mathematics of Emerging Biomedical Imaging. Mathématiques et Applications, vol. 62. Springer, Berlin (2008)
19. Ammari, H., Kang, H.: High-order terms in the asymptotic expansions of the steady-state voltage potentials in the presence of conductivity inhomogeneities of small diameter. SIAM J. Math. Anal. **34**, 1152–1166 (2003)
20. Ammari, H., Kang, H.: Boundary layer techniques for solving the Helmholtz equation in the presence of small inhomogeneities. J. Math. Anal. Appl. **296**, 190–208 (2004)
21. Ammari, H., Kang, H.: Reconstruction of Small Inhomogeneities from Boundary Measurements. Lecture Notes in Mathematics. Springer, Berlin (2004)
22. Ammari, H., Kang, H.: Polarization and Moment Tensors with Applications to Inverse Problems and Effective Medium Theory. Applied Mathematical Sciences, vol. 162. Springer, Berlin (2007)
23. Ammari, H., Vogelius, M.S., Volkov, D.: Asymptotic formulas for perturbations in the electromagnetic fields due to the presence of inhomogeneities of small diameter II. The full Maxwell equations. J. Math. Pures Appl. **80**, 769–814 (2001)
24. Ammari, H., Kang, H., Nakamura, G., Tanuma, K.: Complete asymptotic expansions of solutions of the system of elastostatics in the presence of an inclusion of small diameter and detection of an inclusion. J. Elast. **67**, 97–129 (2002)
25. Ammari, H., Moskow, S., Vogelius, M.: Boundary integral formulas for the reconstruction of electromagnetic imperfections of small diameter. ESAIM Control Optim. Calc. Var. **9**, 49–66 (2003)
26. Ammari, H., Griesmaier, R., Hanke, M.: Identification of small inhomogeneities: asymptotic factorization. Math. Comput. **76**, 1425–1448 (2007)
27. Ammari, H., Bretin, E., Garnier, J., Kang, H., Lee, H., Wahab, A.: Mathematical Methods in Elasticity Imaging. Princeton Series in Applied Mathematics. Princeton University Press, Princeton (2015)
28. Amrouche, C., Bonzom, F.: Exterior problems in the half-space for the Laplace operator in weighted Sobolev spaces. J. Differ. Equ. **246**, 1894–1920 (2009)
29. Amrouche, C., Nečasová, S.: Laplace equation in the half-space with a nonhomogeneous Dirichlet boundary condition. Math. Boh. **126**, 265–274 (2001)
30. Amrouche, C., Girault, V., Giroire, J.: Weighted Sobolev spaces for Laplace's equation in $\mathbb{R}^n$. J. Math. Pures Appl. **73**, 579–606 (1994)
31. Amrouche, C., Nečasová, S., Raudin, Y.: Very weak, generalized and strong solutions to the Stokes system in the half-space. J. Differ. Equ. **244**, 887–915 (2008)
32. Amrouche, C., Dambrine, M., Raudin, Y.: An L^p theory of linear elasticity in the half-space. J. Differ. Equ. **253**, 906–932 (2012)
33. Amrouche, C., Meslameni, M., Nečasová, S.: Linearized Navier-Stokes equations in $\mathbb{R}^3$: an approach in weighted Sobolev spaces. Discrete Contin. Dynam. Syst. **7**, 901–916 (2014)
34. Anderson, E.M.: Dynamics of the formation of cone-sheets, ringdykes, and cauldron-subsidences. Proc. R. Soc. Edinb. **56**, 128–157 (1936)
35. Aspri, A., Beretta, E., Mascia, C.: Asymptotic expansion for harmonic functions in the half-space with a pressurized cavity. Math. Meth. Appl. Sci. **39**(10), 2415–2430 (2016)
36. Aspri, A., Beretta, E., Mascia, C.: Analysis of a Mogi-type model describing surface deformations induced by magma chamber embedded in an elastic half-space. J. École Polytech. Math. **4**, 223–255 (2017)
37. Aspri, A., Beretta, E., Rosset, E.: On an elastic model arising from volcanology: an analysis of the direct and inverse problem. J. Differ. Equ. **265**, 6400–6423 (2018)
38. Battaglia, M., Hill, D.P.: Analytical modeling of gravity changes and crustal deformation at volcanoes: the Long Valley caldera, California, case study. Tectonophysics **471**, 45–57 (2009)
39. Battaglia, M., Segall, P., Murray, J., Cervelli, P., Langbein, J.: The mechanics of unrest at Long Valley caldera, California. 1. Modeling the geometry of the source using GPS, leveling and two-color EDM data. J. Volcanol. Geotherm. Res. **127**, 195–217 (2003)

40. Battaglia, M., Segall, P., Roberts, C.: The mechanics of unrest at Long Valley caldera, California. 2. Constraining the nature of the source using geodetic and micro-gravity data. J. Volcanol. Geotherm. Res. **127**, 219–245 (2003)
41. Battaglia, M., Gottsmann, J., Carbone, D., Fernández, J.: 4D volcano gravimetry. Geophysics **73**, WA3–WA18 (2008)
42. Battaglia, M., Cervelli, P.F., Murray, J.R.: dMODELS: a MATLAB software package for modeling crustal deformation near active faults and volcanic centers. J. Volcanol. Geotherm. Res. **254**, 1–4 (2013)
43. Battaglia, M., Cervelli, P.F., Murray, J.R.: Modeling crustal deformation near active faults and volcanic centers - a catalog of deformation models. U.S. Geological Survey Techniques and Methods, p. 13-B1, 75 pp. (2013)
44. Battaglia, M., Lisowski, M., Dzurisin, D., Poland, M.P., Schilling, S., Diefenbach, A., Wynn, J.: Mass Addition at Mount St. Helens, Washington, Inferred From Repeated Gravity Surveys. J. Geophys. Res.: Solid Earth **123** (2018). https://doi.org/10.1002/2017JB014990
45. Beretta, E., Francini, E.: An asymptotic formula for the displacement field in the presence of thin elastic inhomogeneities. SIAM J. Math. Anal. **38**, 1249–1261 (2006)
46. Beretta, E., Vessella, S.: Stable determination of boundaries from Cauchy data. SIAM J. Math. Anal. **30**, 220–232 (1998)
47. Beretta, E., Mukherjee, A., Vogelius, M.S.: Asymptotic formulas for steady state voltage potentials in the presence of conductivity imperfections of small area. Z. Angew. Math. Phys. **52**, 543–572 (2001)
48. Beretta, E., Francini, E., Vogelius, M.S.: Asymptotic formulas for steady state voltage potentials in the presence of thin inhomogeneities. A rigorous error analysis. J. Math. Pures Appl. **82**, 1277–1301 (2003)
49. Beretta, E., Francini, E., Vessella, S.: Determination of a linear crack in an elastic body from boundary measurements-Lipschitz stability. SIAM J. Math. Anal. **40**, 984–1002 (2008)
50. Berrino, G., Rymer, H., Brown, G.C., Corrado, G.: Gravity-height correlations for unrest at calderas. J. Volcanol. Geotherm. Res. **53**, 11–26 (1992)
51. Bonaccorso, A., Davis, P.M.: Models of ground deformation from vertical volcanic conduits with application to eruptions of Mount St. Helens and Mount Etna. J. Geophys. Res. **104**(B5), 10531–10542 (1999)
52. Bonafede, M.: Hot fluid migration: an efficient source of ground deformation: application to the 1982–1985 crisis at Campi Flegrei-Italy. J. Volcanol. Geotherm. Res. **48**, 187–198 (1991)
53. Brezis, H.: Functional Analysis, Sobolev Spaces and Partial Differential Equations. Springer, New York (2011)
54. Brühl, M., Hanke, M., Vogelius, M.S.: A direct impedance tomography algorithm for locating small inhomogeneities. Numer. Math. **93**, 635–654 (2003)
55. Calderón, A.P.: Uniqueness in the Cauchy problem for partial differential equations. Am. J. Math. **80**, 16–36 (1958)
56. Calderón, A.P.: Cauchy integrals on Lipschitz curves and related operators. Proc. Natl. Acad. Sci. U.S.A. **74**, 1324–1327 (1977)
57. Calderón, A.P., Zygmund, A.: On the existence of certain singular integrals. Acta Math. **88**, 85–139 (1952)
58. Calderón, A.P., Zygmund, A.: On singular integrals. Am. J. Math. **78**, 289–309 (1956)
59. Capdeboscq, Y., Vogelius, M.S.: A general representation formula for boundary voltage perturbations caused by internal conductivity inhomogeneities of low volume fraction. Math. Model. Numer. Anal. **37**, 159–173 (2003)
60. Capdeboscq, Y., Vogelius, M.S.: Optimal asymptotic estimates for the volume of internal inhomogeneities in terms of multiple boundary measurements. Math. Model. Numer. Anal. **3**, 227–240 (2003)
61. Cedio-Fengya, D.J., Moskow, S., Vogelius, M.S.: Identification of conductivity imperfections of small diameter by boundary measurements. Continuous dependence and computational reconstruction. Inverse Prob. **14**, 553–595 (1998)
62. Ciarlet, P.G.: Mathematical Elasticity. Vol. I. Three-Dimensional Elasticity. Studies in Mathematics and Its Applications, vol. 20. North-Holland Publishing Co., Amsterdam (1988)

63. Coifman, R.R., McIntosh, A., Meyer, Y.: L'intégrale de Cauchy définit un opérateur borné sur L^2 pour les courbes lipschitziennes. Ann. Math. **116**, 361–387 (1982)
64. Currenti, G., Del Negro, C., Ganci, G.: Finite element modeling of ground deformation and gravity field at Mt. Etna. Ann. Geophys. **51**(1), 105–119 (2008)
65. Currenti, G., Bonaccorso, A., Del Negro, C., Scandura, D., Boschi, E.: Elasto-plastic modeling of volcano ground deformation. Earth Planet. Sci. Lett. **296**, 311–318 (2010)
66. Dahlberg, B.E., Kenig, C.E., Verchota, G.C.: Boundary value problem for the system of elastostatics in Lipschitz domains. Duke Math. J. **57**, 795–818 (1988)
67. Davis, P.M.: Surface deformation associated with a dipping hydrofracture. J. Geophys. Res. **88**, 5829–5834 (1983)
68. Davis, P.M.: Surface deformation due to inflation of an arbitrarily oriented triaxial ellipsoidal cavity in an elastic half-space, with reference to Kilauea Volcano, Hawaii. J. Geophys. Res. **91**(B7), 7429–7438 (1986)
69. De Natale, G., Pingue, F.: Ground deformation modeling in volcanic areas. In: Scarpa, R., Tilling, R.I. (eds.) Monitoring and Mitigation of Volcano Hazards, pp. 365–388. Springer, Berlin (1996)
70. Di Cristo, M., Rondi, L.: Examples of exponential instability for inverse inclusion and scattering problems. Inverse Prob. **19**, 685–701 (2003)
71. Di Traglia, F., Battaglia, M., Nolesini, T., Lagomarsino, D., Casagli, N.: Shifts in the eruptive styles at Stromboli in 2010–2014 revealed by ground-based InSAR data. Sci. Rep. **5**, 1–11 (2015)
72. Dzurisin, D.: A comprehensive approach to monitoring volcano deformation as a window on the eruption cycle. Rev. Geophys. **41**, 29 pp. (2003)
73. Dzurisin, D.: Volcano Deformation. Geodetic Monitoring Techniques. Springer Praxis Books. Springer, Berlin (2006)
74. Dzurisin, D., Savage, J., Fournier, B.: Recent crustal subsidence at Yellowstone Caldera, Wyoming. Bull. Volcanol. **52**, 247–270 (1990)
75. Escauriaza, L., Fabes, E.B., Verchota, G.: On a regularity theorem for weak solutions to transmission problems with internal Lipschitz boundaries. Proc. Am. Math. Soc. **115**, 1069–1076 (1992)
76. Evans, L.C.: Partial Differential Equations. American Mathematical Society, Providence (1998)
77. Fabes, E.B., Jodeit, M., Lewis, J.E.: Double layer potentials for domains with corners and edges. Indiana Univ. Math. J. **26**, 95–114 (1977)
78. Fabes, E.B., Jodeit, M., Riviére, N.M.: Potential techniques for boundary value problems on C^1 domains. Acta Math. **141**, 165–186 (1978)
79. Fabes, E.B., Kenig, C.E., Verchota, G.: The Dirichlet problem for the Stokes system on Lipschitz domains. Duke Math. J. **57**, 795–818 (1988)
80. Fernández, J., Charco, M., Tiampo, K.F., Jentzsch, G., Rundle, J.B.: Joint interpretation of displacement and gravity data in volcanic area. A test example: Long Valley Caldera, California. Geophys. Res. Lett. **28**, 1063–1066 (2001)
81. Fernández, J., Tiampo, K.F., Jentzsch, G., Charco, M., Rundle, J.B.: Inflation or Deflation? New results for Mayon Volcano applying elastic-gravitational modeling. Geophys. Res. Lett. **28**, 2349–2352 (2001)
82. Fialko, Y., Khazan, Y., Simons, M.: Deformation due to a pressurized horizontal circular crack in an elastic half-space, with applications to volcano geodesy. Geophys. J. Int. **146**(1), 181–190 (2001)
83. Fichera, G.: Sull'esistenza e sul calcolo delle soluzioni dei problemi al contorno, relativi all'equilibrio di un corpo elastico. Annali della Scuola Normale Superiore di Pisa, Classe di Scienze 3e serie **4**, 35–99 (1950)
84. Folland, G.B.: Introduction to Partial Differential Equations. Princeton University Press, Princeton (1995)
85. Friedman, A., Vogelius, M.: Identification of small inhomogeneities of extreme conductivity by boundary measurements: a theorem on continuous dependence. Arch. Ration. Mech. Anal. **105**, 299–326 (1984)

86. Garofalo, N., Lin, F.: Monotonicity properties of variational integrals, A_p weights and unique continuation. Indiana Univ. Math. J. **35**, 245–268 (1986)
87. Gurtin, M.E.: The linear theory of elasticity. In: Encyclopedia of Physics, vol. VI a/2. Springer, Berlin (1972)
88. Guzina, B.B., Bonnet, M.: Topological derivative for the inverse scattering of elastic waves. Q. J. Mech. Appl. Math. **57**, 161–179 (2004)
89. Hadamard, J.: Sur les problemes aux derivées partielles et leur signification physique. Bull. Univ. Princeton **13**, 49–52 (1902)
90. Hadamard, J.: Lectures on Cauchy's Problem in Linear Partial Differential Equations. Dover, New York (1953)
91. Hanouzet, B.: Espaces de Sobolev avec poids. Application au problème de Dirichlet dans un demi-espace. Rendiconti del Seminario Matematico della Università di Padova **46**, 227–272 (1971)
92. Hautmann, S., Gottsmann, J., Sparks, R.S.J., Mattioli, G.S., Sacks, I.S., Strutt, M.H.: Effect of mechanical heterogeneity in arc crust on volcano deformation with application to Soufrière Hills volcano, Monteserrat, West Indies. J. Geophys. Res. Solid Earth **115**, B09203 (2010)
93. Hein Hoernig, R.O.: Green's functions and integral equations for the Laplace and Helmholtz operators in impedance half-spaces. Mathematics, Ecole Polytechnique X (2010)
94. Hickey, J., Gottsmann, L., Mothes, P., Odbert, H., Prutkin, I., Vajda, P.: The ups and downs of volcanic unrest: insights from integrated geodesy and numerical modelling. In: Advances in Volcanology. Springer, Berlin (2017)
95. Isakov, V.: Uniqueness and stability in multidimensional inverse problems. Inverse Prob. **9**, 579–621 (1993)
96. John, F.: Numerical solution of the equation of heat conduction for preceding times. Ann. Mat. Pura Appl. **40**, 129–142 (1955)
97. John, F.: Continuous dependence on data for solutions of partial differential equations with a prescribed bound. Commun. Pure Appl. Math. **13**, 551–585 (1960)
98. Kang, H., Seo, J.K.: The layer potential technique for the inverse conductivity problem. Inverse Prob. **12**, 267–278 (1996)
99. Kato, T.: Perturbation Theory for Linear Operators. Springer, Berlin (1995)
100. Kellogg, O.D.: Foundations of Potential Theory. Springer, Berlin (1967)
101. Kesavan, S.: Topics in Functional Analysis and Applications. New Age International (P) Limited, Publishers, New Delhi (1989)
102. Kondrat'ev, V.A., Oleinik, O.A.: Boundary-value problems for the system of elasticity theory in unbounded domains. Korn's inequalities. Russ. Math. Surv. **43**, 65–119 (1988)
103. Kress, R.: Linear Integral Equations. Springer, Berlin (1989)
104. Kupradze, V.D.: Potential Methods in the Theory of Elasticity. Israel Program for Scientific Translations, Jerusalem (1965)
105. Landis, E.M.: A three-sphere theorem. Dokl. Akad. Nauk SSSR **148**, 277–279 (1963); Engl. trans. Soviet Math. Dokl. **4**, 76–78 (1963)
106. Lewiński, T., Sokołowski, J.: Energy change due to the appearance of cavities in elastic solids. Int. J. Solids Struct. **40**, 1765–1803 (2003)
107. Lisowski, M.: Analytical volcano deformation source models. In: Dzurisin, D. (ed.) Volcano Deformation. Geodetic Monitoring Techniques. Springer Praxis Books, pp. 279–304. Springer, Berlin (2006)
108. Manconi, A., Walter, T.R., Amelung, F.: Effects of mechanical layering on volcano deformation Geophys. J. Int. **170**(2), 952–958 (2007)
109. Masmoudi, N.: About the Hardy inequality. In: Schleicher, D., Lackmann, M. (eds.) Invitation to Mathematics. From Competitions to Research, pp. 165–180. Springer, Berlin (2011)
110. McLean, W.C.: Strongly Elliptic Systems and Boundary Integral Equations. Cambridge University Press, Cambridge (2000)
111. McTigue, D.F.: Elastic stress and deformation near a finite spherical magma body: resolution of the point source paradox. J. Geophys. Res. **92**, 12931–12940 (1987)

112. Mindlin, R.D.: Force at a point in the interior of a semiinfinite solid. J. Appl. Phys. **7**, 195–202 (1936)
113. Mindlin, R.D.: Force at a point in the interior of a semi-infinite solid. In: Proceedings of the First Midwestern Conference on Solid Mechanics, April, University of Illinois, Urbana (1954)
114. Mogi, K.: Relations between the eruptions of various volcanoes and the deformation of the ground surfaces around them. Bull. Earthq. Res. Inst. Univ. Tokyo **36**, 99–134 (1958)
115. Morassi, A., Rosset, E.: Stable determination of cavities in elastic bodies. Inverse Prob. **20**, 453–480 (2004)
116. Morassi, A., Rosset, E.: Uniqueness and stability in determining a rigid inclusion in an elastic body. Mem. Am. Math. Soc. **200**(938), 453–480 (2009)
117. Morassi, A., Rosset, E.: Stable determination of an inclusion in an inhomogeneous elastic body by boundary measurements. Rend. Istit. Mat. Univ. Trieste **48**, 101–120 (2016)
118. Morassi, A., Rosset, E., Vessella, S.: Detecting general inclusions in elastic plates Inverse Prob. **25**, 045009 (14 pp.) (2009)
119. Morassi, A., Rosset, E., Vessella, S.: Stable determination of a rigid inclusion in an anisotropic elastic plate. SIAM J. Math. Anal. **44**, 2204–2235 (2012)
120. Morassi, A., Rosset, E., Vessella, S.: Recent results about the detection of unknown boundaries and inclusions in elastic plates. J. Inverse Ill-Posed Probl. **21**, 311–352 (2013)
121. Okada, Y.: Surface deformation due to shear and tensile faults in a half-space. Bull. Seismol. Soc. Am. **75**, 1135–1154 (1985)
122. Rymer, H.: Microgravity change as a precursor to volcanic activity. J. Volcanol. Geotherm. Res. **61**, 311–328 (1994)
123. Rynne, B.P., Youngson, M.A.: Linear Functional Analysis. Springer, Berlin (2008)
124. Salsa, S.: Equazioni a derivate parziali metodi, modelli e applicazioni. Springer, Berlin (2010)
125. Segall, P.: Earthquake and Volcano Deformation. Princeton University Press, Princeton (2010)
126. Segall, P.: Volcano deformation and eruption forecasting. Geol. Soc. Spec. Publ. **380**(1), 85–106 (2013)
127. Segall, P.: Magma chambers: what we can, and cannot, learn from volcano geodesy. Philos. Trans. R. Soc. A **377**, 20180158 (2019)
128. Tizzani, P., Battaglia, M., Castaldo, R., Pepe, A., Zeni, G., Lanari, R.: Magma and fluid migration at Yellowstone Caldera in the last three decades inferred from InSAR, leveling, and gravity measurements. J. Geophys. Res.: Solid Earth **120**, 2627–2647 (2015)
129. Trasatti, E., Giunchi, C., Bonafede, M.: Effects of topography and rheological layering on ground deformation in volcanic regions. J. Volcanol. Geotherm. Res. **122**, 89–110 (2003)
130. Valent, T.: Boundary Value Problems of Finite Elasticity, Local Theorems on Existence, Uniqueness and Analytic Dependence on Data. Springer, New York (1988)
131. Verchota, G.C.: Layer potentials and regularity for the Dirichlet problem for Laplace's equation in Lipschitz domains. J. Funct. Anal. **59**, 572–611 (1984)
132. Vogelius, M.S., Volkov, D.: Asymptotic formulas for perturbations in the electromagnetic fields due to the presence of inhomogeneities. M2AN Math. Model. Numer. Anal. **34**, 723–748 (2000)
133. Volpert, V.: Elliptic Partial Differential Equations. Volume 1: Fredholm Theory of Elliptic Problems in Unbounded Domains. Springer, Basel (2011)
134. Wicks, C., Thatcher, W., Dzurisin, D.: Migration of fluids beneath Yellowstone Caldera inferred from satellite radar interferometry. Science **282**, 458–462 (1998)
135. Williams, C.A., Wadge, G.: The effects of topography on magma chamber deformation models: application to Mt. Etna and radar interferometry. Geophys. Res. Lett. **25**(10), 1549–1552 (1998)
136. Yamakawa, N.: On the strain produced in a semi-infinite elastic solid by an interior source of stress. Zisin (J. Seismol. Soc. Jpn.) **8**, 84–98 (1955)
137. Yang, X.-M., Davis, P.M., Dieterich, J.H.: Deformation from inflation of a dipping finite prolate spheroid in an elastic half-space as a model for volcanic stressing. J. Geophys. Res. Solid Earth **93**(B5), 4249–4257 (1988)
138. Yosida, K.: Functional Analysis. Springer, Berlin (1980)

Index

A. Aspri, *An Elastic Model for Volcanology*, Lecture Notes in Geosystems Mathematics and Computing, https://doi.org/10.1007/978-3-030-31475-0

T

W

Printed in the United States
by Baker & Taylor Publisher Services